T0296856

The Physics of Computing

The Physics of Computing

Marilyn Wolf

AMSTERDAM • BOSTON • HEIDELBERG • LONDON
NEW YORK • OXFORD • PARIS • SAN DIEGO
SAN FRANCISCO • SINGAPORE • SYDNEY • TOKYO

Morgan Kaufmann is an imprint of Elsevier

Morgan Kaufmann is an imprint of Elsevier
50 Hampshire Street, 5th Floor, Cambridge, MA 02139, United States

Copyright © 2017 Elsevier Inc. All rights reserved.

No part of this publication may be reproduced or transmitted in any form or by any means, electronic or mechanical, including photocopying, recording, or any information storage and retrieval system, without permission in writing from the publisher. Details on how to seek permission, further information about the Publisher's permissions policies and our arrangements with organizations such as the Copyright Clearance Center and the Copyright Licensing Agency, can be found at our website: www.elsevier.com/permissions.

This book and the individual contributions contained in it are protected under copyright by the Publisher (other than as may be noted herein).

Notices

Knowledge and best practice in this field are constantly changing. As new research and experience broaden our understanding, changes in research methods, professional practices, or medical treatment may become necessary.

Practitioners and researchers must always rely on their own experience and knowledge in evaluating and using any information, methods, compounds, or experiments described herein. In using such information or methods they should be mindful of their own safety and the safety of others, including parties for whom they have a professional responsibility.

To the fullest extent of the law, neither the Publisher nor the authors, contributors, or editors, assume any liability for any injury and/or damage to persons or property as a matter of products liability, negligence or otherwise, or from any use or operation of any methods, products, instructions, or ideas contained in the material herein.

Library of Congress Cataloging-in-Publication Data
A catalog record for this book is available from the Library of Congress

British Library Cataloguing-in-Publication Data
A catalogue record for this book is available from the British Library

ISBN: 978-0-12-809381-8

For information on all Morgan Kaufmann publications
visit our website at https://www.elsevier.com/

www.elsevier.com • www.bookaid.org

Working together
to grow libraries in
developing countries

Publisher: Todd Green
Acquisition Editor: Steve Merken
Editorial Project Manager: Nate McFadden
Production Project Manager: Mohanambal Natarajan
Cover Designer: Matthew Limbert

Typeset by TNQ Books and Journals

Contents

Preface ix

CHAPTER 1 Electronic Computers 1
  1.1 Introduction 1
  1.2 The long road to computers 1
        1.2.1 Mechanical computing devices 2
        1.2.2 Theories of computing 4
        1.2.3 Electronic computers 6
  1.3 Computer system metrics 9
  1.4 A tour of this book 10
  1.5 Synthesis 11
  Questions 11

CHAPTER 2 Transistors and Integrated Circuits 13
  2.1 Introduction 13
  2.2 Electron devices and electronic circuits 13
        2.2.1 Early vacuum tube devices 13
        2.2.2 Vacuum tube triode 15
  2.3 Physics of materials 18
        2.3.1 Metals 19
        2.3.2 Boltzmann's constant and temperature 22
        2.3.3 Semiconductors 24
  2.4 Solid-state devices 29
        2.4.1 Semiconductor diode 30
        2.4.2 MOS capacitor 33
        2.4.3 Basic MOSFET operation 37
        2.4.4 Advanced MOSFET characteristics 45
  2.5 Integrated circuits 48
        2.5.1 Moore's Law 49
        2.5.2 Manufacturing processes 51
        2.5.3 Lithography 55
        2.5.4 Yield 57
        2.5.5 Separation of concerns 58
  2.6 Synthesis 59
  Questions 59

CHAPTER 3 Logic Gates 63
  3.1 Introduction 63
  3.2 The CMOS inverter 64

3.3 Static gate characteristics 67
3.4 Delay 72
    3.4.1 Transistor models 73
    3.4.2 RC models for delay 76
    3.4.3 Drive and loads.. 83
3.5 Power and energy 85
3.6 Scaling theory 90
3.7 Reliability.. 94
3.8 Synthesis 96
Questions... 96

CHAPTER 4 **Sequential Machines** **99**
4.1 Introduction... 99
4.2 Combinational logic 99
    4.2.1 The event model 99
    4.2.2 The network model 100
    4.2.3 Gain and reliability.. 103
    4.2.4 Gain and delay... 105
    4.2.5 Delay and power... 108
    4.2.6 Noise and reliability in logic and interconnect 109
    4.2.7 Power supply and reliability 109
    4.2.8 Noise and input/output coupling........................... 115
4.3 Interconnect 116
    4.3.1 Parasitic impedance.. 116
    4.3.2 Transmission lines 118
    4.3.3 Crosstalk 124
    4.3.4 Wiring complexity and Rent's Rule 126
4.4 Sequential machines.. 128
    4.4.1 Sequential models 128
    4.4.2 Registers 131
    4.4.3 Clocking 134
    4.4.4 Metastability 140
4.5 Synthesis 144
Questions... 145

CHAPTER 5 **Processors and Systems**................................... **149**
5.1 Introduction... 149
5.2 System reliability 150
5.3 Processors 153
    5.3.1 Microprocessor characteristics............................. 153
    5.3.2 Busses and interconnect 155
    5.3.3 Global communication 159
    5.3.4 Clocking 161

5.4 Memory 168
  5.4.1 Memory structures.. 168
  5.4.2 Memory system performance............................. 172
  5.4.3 DRAM systems 174
  5.4.4 DRAM reliability 176
5.5 Mass storage 177
  5.5.1 Magnetic disk drives 177
  5.5.2 Flash memory 178
  5.5.3 Storage and performance 181
5.6 System power consumption.. 182
  5.6.1 Server systems... 182
  5.6.2 Mobile systems and batteries........................... 185
  5.6.3 Power management 188
5.7 Heat transfer 189
  5.7.1 Heat transfer characteristics............................. 190
  5.7.2 Heat transfer modeling..................................... 192
  5.7.3 Heat and reliability.. 198
  5.7.4 Thermal management.. 200
5.8 Synthesis 201
Questions.. 201

CHAPTER 6   Input and Output **205**
6.1 Introduction.. 205
6.2 Displays.. 205
6.3 Image sensors... 211
6.4 Touch sensors... 216
6.5 Microphones 217
6.6 Accelerometers and inertial sensors............................... 218
6.7 Synthesis 220
Questions.. 220

CHAPTER 7   Emerging Technologies **221**
7.1 Introduction.. 221
7.2 Carbon nanotubes 221
  7.2.1 Nanotube transistors... 222
7.3 Quantum computers.. 223
7.4 Synthesis 228

Appendix A: Useful Constants and Formulas 229
Appendix B: Circuits... 233
Appendix C: Probability.. 241
Appendix D: Advanced Topics 243
References... 253
Index 259

Preface

This book is an attempt to take a step back and provide a longer, more foundational view of computer engineering. While a lot of software design proceeded from the theoretical to the experimental—think about sorting, for example—computer system design has long taken an example-driven approach. Studying particular computer systems was useful in the early days when we were not quite sure what major concepts undergirded the design of these systems. But today we are much more capable of making generalized statements about many topics in computer engineering. Those foundational topics should form the core of our approach to computer engineering.

A foundational approach to digital systems is particularly important because it is so hard today to touch and see logic in operation. When I was a student, most systems were boards made of SSI and MSI logic. We had no choice but to deal with circuits and waveforms. Today, all that is hidden inside a chip; even boards have smaller wires and are harder to analyze. Many students today think that 1s and 0s float through chips—they have no conception of voltages and currents in computers.

This book does not ask students to design *anything*. Instead, we walk students through the principles that define the computer system design space. I like to think of the concepts in this course as knobs and meters on a control panel. Changing the settings of knobs change the values on the meters; the twist of one knob may change several meters. Consider, for example, the cascading effects of reducing MOS dielectric thickness: transistor transconductance changes, which in turn affects gate delay for the better and leakage current for the worse.

Engineers do not design systems in a vacuum. Instead, they design to a set of goals or requirements. The traditional goal for computer architects is performance, or more precisely throughput. But computers are actually designed with several other goals in mind, with energy/power and reliability paramount. Performance, energy, and reliability all have their roots in physical phenomenon. And they are inextricably linked by physics: improving one metric incurs costs in the other metrics. As in the rest of life, there is no free lunch in computer design.

Computer engineering is a relatively young field. Introductory courses have concentrated on building and taxonomy—build something, see the different ways to build something. As the field matures, it is time to start thinking about new ways to present the basics. The biologist E.O. Wilson said, "A field is initially defined by the questions it asks and eventually defined by the answers it provides." After 70 years of electronic computer engineering, it is time to start thinking of the field in terms of answers.

I have been thinking about computer design for a long time. Performance was for many years the paramount metric for computer design, although the ASIC world was always interested in area as a proxy for cost. I started to think more seriously about other fundamental limits, particularly power, for the class on pervasive information systems taught by Perry Cook and myself. A new course at Georgia Tech gave me the impetus to take this train of thought to its logical conclusion.

This course takes a broad view of both computer engineering and physics. Some of the most basic phenomena in computer architecture and even software design—the memory wall, the power wall, the race to dark—are due to fundamental physics. But the physics required to understand these major computational issues is much more than classical Shockley semiconductor theory. We need to understand thermodynamics, electrostatics, and good deal of circuit theory.

As I thought about this material, I came to regard Boltzmann's constant is a key concept. k pops up everywhere: the diode equation, Arrhenius's equation, temperature, and the list goes on. Boltzmann's constant links temperature and energy, so it should not be a surprise that it is so tightly woven into the topics of this book.

Some of the material in this book is unique to modern CMOS—leakage mechanisms. But some of the topics can be applied to a wide range of circuit and device technologies. Delay through logic networks, metastability, and the fundamentals of reliability are examples of foundational concepts that computer engineers should understand even if CMOS is swept off the map by some other technology.

Parts of this book will probably strike some readers as hopelessly simplified and compressed. I hope that those same readers will find other parts of the book dense and hard to follow. The only way to understand how computers work is to understand the relationships between an entire collection of topics that do not always seem at first glance to be related. We know those knobs on the design process are linked because a change to one parameter that we think will help our design often results in side effects that negate much of the benefit we sought. I worked very hard to find the simplest possible description for a lot of concepts. Hopefully they are presented in a way that gives a basic understanding. The reader who is interested in a deeper understanding of one of these topics can pursue each one in more depth. But the singular goal of this book is to provide a unified description of the fundamental physical principles of computing machines.

Both computer engineers and electrical engineers are potential audiences for this book. These two groups come to the material with very different backgrounds. Computer engineers often have limited experience in circuit design. While they may have some knowledge of ideas like Kirchhoff's laws, they usually are not extremely comfortable with circuit analysis. Electrical engineers may not always have a detailed understanding of computer architecture. One of my challenges in writing this book was to provide enough background for each audience without overloading them.

I have mentioned a number of historical discoveries and inventions for several reasons. First, going through the sequence of ideas that led to today's design practices helps both to remind us that there is often more than one way to do things and to highlight the true advantages of the methods that were settled upon by history. Second, semiconductor physics and computer engineering have produced some of the most important inventions of the 20th century. These inventions will be used for centuries to come. We should not let ourselves become so accustomed to these ideas that we lose all capacity to marvel at them and at their inventors.

Richard Feynman's *Lectures on Computing* was an early inspiration for this class, but much of that book is devoted to quantum computing. That book does not consider many topics in traditional computing. For example, metastability is a fundamental physical concept in computer system design that Feynman does not touch. We salute him for his early recognition of the physical nature of computing. We also thank him for the *Lectures on Physics*, which provided a clear, concise way to think about many of the basic physical phenomena that underlie computing.

Several tips of the hat are in order. My friend and colleague Saibal Mukhopodhyay proposed the *Physical Foundations of Computer Engineering* course and provided many concrete suggestions, particularly on reliability and leakage. My thanks go out to him for his patience and insight. Dave Coelho graciously provided information on his power distribution system. Kees Vissers invented the welder current comparison. Alec Ishii advised me on clock distribution. Kevin Cao suggested how to make best use of the Predictive Technology Models. Bruce Jacob gave insight into DRAM. Srini Devadas's suggestions led to the advanced topics appendix. Tom Conte provided the core memory and Pentium Pro as well as many discussions on the present state and future of computing. Thanks to the reviewers for their helpful comments. And many thanks to my editor Nate McFadden for guiding this book through the long development and production process. Any faults in this book can be traced entirely and solely to me.

Marilyn Wolf
Atlanta, GA

CHAPTER

Electronic Computers

1

1.1 Introduction

We take computers for granted. But what is a computer? The computers we use every day can be summarized as **EBT**: *electronic, binary, Turing machines.* This combination of technologies has allowed us to conquer problems as diverse as weather forecasting, self-driving vehicles, and shopping.

But the path that led us to today's computers is long, and the EBT approach to computing is surprisingly recent. Some of the history helps us understand why we build computers the way we do. Machines that could compute—calculate with numbers and perform other multistep tasks—were originally conceived of and built with the only devices available, namely mechanical devices. These mechanical computers had severe limitations that were overcome only with the creation of electronic devices.

We will start with a brief introduction to the history of computing machines. We will then discuss the theory of computing that was developed based on this early experience and which underpins our modern ideas of computers. We then discuss the metrics we use to evaluate computer designs and the trade-offs between these goals that physics imposes on us. We will wrap up with a survey of the remainder of this book.

1.2 The long road to computers

Mechanical computation was inspired in large part by the need to control machines. People could not react quickly and reliably enough to maintain the proper operation of the equipment that appeared during the Industrial Revolution. Early mechanical computing devices were analog—they operated on continuous values. But machines that manipulated discrete values arrived surprisingly early. Eventually, theory was developed to describe these types of machines. One of those theories, the *Turing machine*, is the blueprint for the way in which we design computers today.

The Physics of Computing. http://dx.doi.org/10.1016/B978-0-12-809381-8.00001-8
Copyright © 2017 Elsevier Inc. All rights reserved.

1.2.1 **Mechanical computing devices**

Mechanical devices are much older than electrical devices. The subtlety required to make machines with complex behaviors should not be minimized. Those early inventors lacked hardened metals with which to make both the machines themselves and the tools required to shape the parts. They also lacked machine tools that we take for granted today. Even relatively simple machines needed to be controlled without human intervention. The invention of the steam engine spurred demand for automatic control of machines.

Control

The **governor** is a prime example of an early analog computer built from mechanical components. James Watt, the inventor of the first practical steam engine, designed a governor to control the speed of the steam engine, and similar devices had been used before. He built his governor in 1788. Fig. 1.1 shows the operation of Watt's governor. The central shaft is rotated and is connected to the steam engine's crankshaft. Two shafts are hinged at the top, with a weight at the end of each. A set of levers connect the weighted shafts to the engine's throttle—the higher the weights, the more closed the throttle. As the central shaft (and the engine's crankshaft) rotates at higher speed, the weights lift themselves up due to centrifugal force, causing the engine to be throttled down. This mechanism implements **negative feedback**, a fundamental principle in control.

Memory

The Jacquard loom demonstrates an important concept, built with mechanical elements, which would be fundamental to our modern conception of computers: **discrete memory**. The loom, which was created in 1801, was invented to automate the weaving of complex patterns in fabrics. It created a huge mass market for sophisticated fabrics that previously had been available only to the very wealthy. As Fig. 1.2 shows, a fabric consists of threads in both x, known as the weft thread, and the warp thread in y. The loom's shuttle moves the warp thread back and forth. Lifting the weft threads with small hooks known as bolus hooks selectively causes the warp thread to be either above or below the weft, selecting when it will be visible. The more complex the pattern, the more complex the patterns of thread liftings; a complex pattern could lift hundreds of threads in a different pattern on each row. The Jacquard loom stored that complex pattern in punched card: a hole in the card allowed the bolus hook to

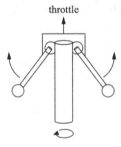

FIGURE 1.1

A mechanical governor.

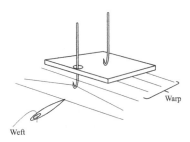

FIGURE 1.2

Operation of a Jacquard loom.

descend and grab its weft thread; no hole meant that the weft was not lifted at that point. Not only could the card represent complex two-dimensional patterns, but the entire fabric's pattern could be changed simply by replacing the card.

Babbage machines   Charles Babbage started work in 1822 on a **difference machine** to calculate numbers, specifically to compute polynomial functions. His work was motivated by the need to tables of functions that were more accurate than those compiled by hand calculation, an error-prone process. At the heart of this machine's operation is the use of gears to count; this principle was also used in the mechanical calculators that were widely used well into the second half of the 20th century. The basic principle behind mechanical counting is simple and elegant, as shown in Fig. 1.3. We use a shaft whose position represents the value of a digit in the counting sum. We can put a radial mark on the end of each shaft, then a dial around the shaft with a position for each digit; the mark points the value of the shaft's digit. To build a two-digit counter, we connect the two shafts with gears. In this figure, the small gear has 5 teeth and the large gear has 15 teeth, giving a gearing ratio of 3:1. Three full rotations of the small gear produce one complete rotation of the large gear. The large gear remembers the number of teeth turned against it by the small gear up to a limit of 15. If we want to count in base 10, we gear the shafts at a 10:1 ratio: 10 rotations of the ones-digit

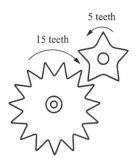

FIGURE 1.3

Using gears to count.

shaft result in one rotation of the tens-digit shaft. Now, as we turn the ones-digit shaft, we can count either up or down.

Babbage later went on to design an **analytical engine** that had the essential elements of a modern computer but in mechanical form: input and output in the form of punched cards (and even a plotter for output); a program memory that could perform what we now know as branches and loops. But he never finished the construction of either his difference machine or the analytical engine.

Hollerith punch cards

Herman Hollerith combined the idea of punched cards as memory with counting machines. His tabulating machine [Hol88] punched holes at various locations in paper cards to record data; the cards were similar to the cards of the Jacquard loom. He then scanned decks of cards through electromechanical readers that counted the numbers of holes at various locations. His machines were used to tabulate data for the 1890 US Census. His company became one of the building blocks for the modern IBM Corporation.

All these machines were great advances. The Jacquard loom and Hollerith's tabulation machines provided huge increases in productivity for important tasks. But in retrospect, we can see some inherent problems with the use of mechanical devices for computing that would ultimately limit how far this line of development could go:

• Mechanical components are relatively heavy and introduce noticeable friction. Both these properties mean that we have to use substantial amounts of energy to make them work. Even with today's nanometer-scale electronic computers, energy is a prime enemy.

• Mechanical components are also subject to wear, which introduces *slop* in the connections between components. In the case of mechanical analog computers such as the Watt regulator, the accuracy of the control law is limited by the accuracy with which the mechanism can move. In the case of discrete systems such as the Babbage engines, the mechanical components had to be accurately aligned, and the complex organization of these devices meant that tolerances in one part of the machine affect the tolerances elsewhere.

1.2.2 Theories of computing

Mathematical logic

Mathematicians spent several centuries developing mathematical notation and theories that freed them from the imprecision of written language. George Boole introduced an algebra for logical expressions that we now know as **Boolean algebra**. He developed laws for basic logical operations such as AND, OR, and NOT. One simple way to describe a logical function is a **truth table** as shown in Fig. 1.4: the left-hand columns enumerate the possible arguments to the function, and the right-hand column gives the result for each. Boole's mathematics is an algebra because its rules are analogous to those of traditional numerical arithmetic.

Lambda calculus

One of the important questions in mathematics at the start of the 20th century was the nature of computing—what functions could be calculated? Mathematicians developed more than one answer to this question. Their answers turned out to be important not just for theory but also to shape the organization of computing machines.

a	b	AND(a,b)
0	0	0
0	1	0
1	0	0
1	1	1

FIGURE 1.4

A Boolean truth table.

The first important theory of computing was developed by Alonzo Church of Princeton University in the 1930s. His theory used functions known as lambda terms to manipulate variables. This theory forms the basis for modern functional programming; LISP and Prolog are two important languages that were inspired by Church's approach to computing.

Turing machines

However, another theory ultimately proved to be more influential on the design of electronic computers. In 1936, Alan Turing, who was Church's student, published his own theory of computing which we now know as the **Turing machine**. As shown in Fig. 1.5, his formalism included several elements:

- An infinite tape divided into cells that can hold discrete values.
- A head that can read and write cells as well as move the tape in both directions.
- A set of rules in the head that tell it what to do based on the value of the cell currently being read.

Turing proved that his machine was equivalent to Church's lambda calculus—both could compute the same set of functions. But the notion of memory was key to the transformation of theory to machines: the tape corresponds to the computer's memory (of course, physical machines had finite memory); the head corresponds to the CPU; the directions for the head correspond to the program.

Turing's theoretical model of the computing process bears significant similarities to the computers we build today:

- They operate on discrete values.
- They operate in discrete time.
- They read and write values on a separate memory.

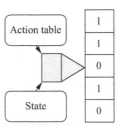

FIGURE 1.5

Organization of a Turing machine.

Highlight 1.1

Turing machines operate on discrete values over discrete time.

Example 1.1: Everything I Need To Know About Mathematics I Learned From Star Trek™

In the original series episode "Wolf in the Fold," the entity Redjac took control of the Enterprise's computer. Spock gave the computer a Class-A compulsory directory to compute π to the last digit. The computer devoted all its resources to the computation, pushing out Redjac. Spock explained the rationale for his command: "the value of π is a transcendental figure without resolution." Or, in computational terms, the procedure to compute π never terminates.

1.2.3 Electronic computers

Binary logic

The design of electronic computers based on vacuum tubes was motivated by the need for scientific calculation for World War II. John Atanasoff and Clifford Berry, working at Iowa State University, built the **electronic computer** shown in Fig. 1.6. Their machine embodied a critical design decision: the use of binary logic. One of

FIGURE 1.6

The Atanasoff-Berry computer.

Courtesy Special Collections/Iowa State University.

the key questions in the design of computers is how to use continuous physical values to represent discrete values. While one could make circuits that manipulated discrete values in several ways, they decided that the most reliable circuits used **binary coding**: logical false or 0 and logical truth or 1. Circuits that tried to represent a larger number of discrete values were more sensitive to noise. This allowed them to use vacuum tube amplifiers to perform logical functions. They represented a logic 0 using the negative power supply voltage and logic 1 as the positive power supply voltage. If one of their logic circuits received an input at one of these values, the output voltages would also be at one of these extremes. If the logic received a degraded voltage, either slightly above the negative supply or below the positive supply, the amplifier would drive it to the saturated value. Amplifying logic gates that push signals toward the power supply rails is known as **saturating logic**. The Atanasoff-Berry computer used a rotating drum to store data but it did not have a writable program store.

At Bell Labs, Claude Shannon developed a theory to optimize networks of switches using Boolean algebra. He also experimented with using switching networks for computation. He built an electromechanical mouse that used a hidden switching network to learn how to run a maze.

The first programmable electronic computer was Colossus, built by Tommy Flowers of the British Post Office Research Station to break German codes. After WWII, work on computers continued and accelerated. The physicist John von Neumann led an effort at the Institute of Advanced Studies in Princeton, New Jersey, to build a digital computer. He also proposed an influential model that we still call the **von Neumann machine**. As shown in Fig. 1.7, his model divided the machine into two distinct pieces: a **central processing unit** (**CPU**) and a separate memory that holds both instructions and data.

Memory

A key problem in early computing was storing values. A variety of technologies were used. Mercury delay lines were commonly used—a bit was represented by a pulse or the lack of a pulse transmitted down the tube. Delay lines act much like what we now know as shift registers: data are entered at one end and come out at the other end after a fixed delay. Reading a particular bit of memory required knowing the current state of the memory and waiting until the desired bit came out. Storage for more than one delay line period required resending the bit through the delay line.

Magnetic core memory was a key innovation because it allowed for random access of all the memory locations. Fig. 1.8 shows a 32×32 bit core module. A bit was stored in the orientation of the magnetic domain of the core. Row (x) and column (y) lines allowed individual cores to be addressed. A shared sense/inhibit line was used to sense the result of a read and to control the result of a write. Reading was a destructive operation—the value was wiped out during the read operation and had to be rewritten. However, the wires through the cores had to be threaded manually, making it impractical for large memories. Semiconductor memory ultimately took over from core memory in large part because semiconductor memory could be manufactured in large volumes.

Analog computers

An alternative technology to digital was analog computation. Analog computers saw more widespread use in the early days of computing and still have some uses

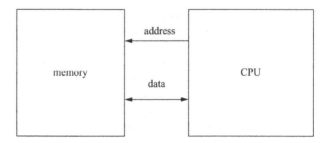

FIGURE 1.7

Organization of a von Neumann machine.

Core memory module

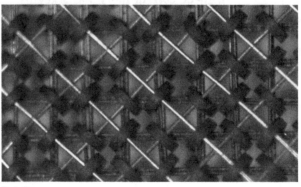

Detail

FIGURE 1.8

A core memory.

FIGURE 1.9

An analog music synthesizer.

Courtesy Moog Music, Inc.

today. A 1958 article describes a room-sized analog computer used to simulate the operation of a DC-8 aircraft [Pre58]; the article explains that the circuits that simulate the engines are on one printed circuit board and that a different type of engine can be installed in the simulator by swapping circuit boards. Analog computers are still used for music synthesis. The Moog synthesizer shown in Fig. 1.9 has circuits to generate and filter signals. The parameters of these circuits are adjusted by the musician using knobs. The circuits can be connected together in different ways using patch cords.

1.3 Computer system metrics

A computer system has to operate in a functionally correct manner but that is not sufficient for most applications—the computer must satisfy a number of other metrics related to its physical characteristics. Several of the most important characteristics of computer systems are directly related to their underlying physical principles:

- *Performance*: how fast does the computer run?
- *Energy and power*: how much energy is required to perform an operation?
- *Thermal*: how hot does the computer run?
- *Reliability*: how frequently does it give the wrong result?

These physical properties are important in themselves to the users of computers. They are particularly important to computer system designers because we must make *trade-offs* between desirable physical properties—there is no free lunch.

A simple example of a design trade-off is performance versus energy. Although gate delay is not the only factor in the execution speed of a processor, it is certainly an important factor. For a given type of gate in a given circuit environment, the only way to make that gate run faster is for it to consume more energy.

Thermal properties have moved to the forefront of concerns for computer designers. Much of the energy consumed by computers is dissipated by heat. Heat can cause catastrophic failure, as evidenced by YouTube videos of CPUs catching fire. But heat is much more likely to cause any number of less extreme failures. Heat causes chips to age and wear out faster.

Reliability is a more subtle but critically important trade-off. Computations rely on transforming the physical state of the machine—the energy of electrons, magnetic domains, etc. A number of physical phenomena can corrupt the result of a computation. We can reduce the effect of those disturbances by using more energy in the system.

These trade-offs are sometimes made within the domain of physics. However, we sometimes solve physics-based problems by other approaches at higher levels of abstraction. For example, thermal and energy problems are both handled in part by operating system mechanisms that monitor thermal and power consumption and adjust the system's operation accordingly.

1.4 **A tour of this book**

This book does not concern itself with the design of any particular computer. It instead looks at the physical principles of computers at several levels of abstraction. Those principles allow us to make particular design decisions. A good understanding of physical principles is particularly important when making design trade-offs.

In the remaining chapters we will study the physics of computing from the bottom up.

- Chapter 2 describes MOS transistors and integrated circuits. The characteristics of transistors are the foundation of logic circuit design. We also need to understand the structure of integrated circuits to understand topics such as interconnect.
- Chapter 3 uses the inverter to describe the physical principles of logic gates. Although we use a number of different logic gates in practice—NAND, NOR, etc.—concentrating on the inverter allows us to explain the basic physical properties of gates.
- Chapter 4 builds on our discussion of single logic gates to understand sequential machines. The delay of a network gates is, unfortunately, not as simple to calculate as adding together the gates in isolation. The properties of registers and clocks are vitally important to our understanding of computer systems.
- Chapter 5 reaches our goal of understanding CPUs and computer systems. We look at the memory wall, the power wall, clock distribution, and storage. Performance, energy, thermal, and reliability all play critical roles in the design of systems.
- Chapter 6 describes a number of input and output devices. Multimedia and the Internet-of-Things drive a great deal of the demand for computer systems. The success of those areas depends on I/O devices such as image sensors, displays, and accelerometers.

- Chapter 7 looks at emerging technologies for computing, including carbon nanotubes and quantum computers.
- The appendices provide reference material. Appendix A summarizes useful constants and formulas. Appendix B covers some basic concepts in electrical circuits. Appendix C describes some basic facts of probability. Appendix D covers a few advanced topics on devices and circuits.

1.5 Synthesis

- Mechanical computers were used to control machines and also for purely computational tasks.
- Early electronic computers used vacuum tubes, which were large, unreliable, and power hungry.
- The physical design of computers is guided by the Turing and von Neumann models: separate memory and processor; a distinction between data and program; discrete values; discrete time.
- The physical properties of computer systems are some of its key characteristics. Performance, energy, thermal, and reliability are all directly determined by the physics of computing.
- Desirable physical characteristics must be traded off against each other in the design of computers.

Questions

Q1-1   Can we build the infinite memory of the Turing machine? What are the effects of a finite memory on the conclusions we can draw about Turing machines?

Q1-2   A machine can move its tape forward but not backward. Is it Turing complete? Explain.

Q1-3   Encode each of these numbers into unsigned binary form. The subscripts give the base of the number.
   (a) 40_{10}
   (b) 40_8
   (c) 100_{10}
   (d) 100_8

Q1-4   Does operating at higher temperatures make a computer less likely or more likely to fail? Explain.

Q1-5   Do computers that burn more power generally run faster? Explain.

Transistors and Integrated Circuits

2

2.1 Introduction

This chapter concentrates on the physics of electron devices and their most prolific embodiment, the integrated circuit. We will start with a brief survey of the history of electron devices and some basic concepts in physics. With that background in mind, we will dive into the characteristics of several electron devices—the MOS capacitor, diode, and MOSFET—as then go onto integrated circuits.

2.2 Electron devices and electronic circuits

Electron devices rely on the specific properties of electron, such as charge. Electrical currents can be easily modeled as fluid flows. In contrast, electron devices such as the transistor rely on much more complex properties and interactions.

The first electron devices were built from vacuum tubes. The limitations of those devices—their bulk, fragility, and power consumption—ultimately led to the invention of their replacement, the transistor.

2.2.1 Early vacuum tube devices

Patent

Amazingly, work on the devices that allowed us to move from mechanical to electronic computation started before the discovery of the electron in 1897. The history of electronic devices started with Thomas Edison. Edison put a great deal of effort into the incandescent lamp, which ran an electric current through a small filament to heat it, resulting in the filament glowing and emitting light; the filament was operated in a vacuum bulb to slow its deterioration. In 1883, he experimented with adding an additional plate inside the bulb and noticed that he could measure a current between the filament and the plate, but only in one direction [Edi84]. As shown in Fig. 2.1, the filament is connected to a battery to heat it while the plate is connected to its own battery. If the battery is connected to the plate with the polarity shown in the figure, a current is conducted; if the battery is connected in the opposite polarity, no current flows. Others named the underlying physical phenomenon the **Edison Effect**.

The Physics of Computing. http://dx.doi.org/10.1016/B978-0-12-809381-8.00002-X
Copyright © 2017 Elsevier Inc. All rights reserved.

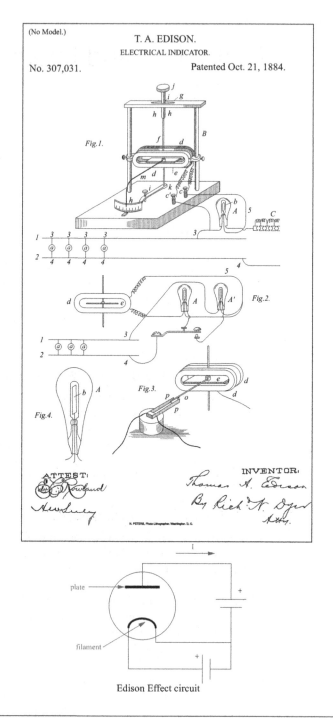

FIGURE 2.1

Operation of an Edison Effect device.

The relationship between heated materials and electron behavior is now known as **thermionics**.

Vacuum tube diode

Motivated by the need for improved radio signal detectors, John Ambrose Fleming created an improved device based upon the Edison Effect [Fle05]. The **Fleming valve**, shown in Fig. 2.2, was the first diode, which had two terminals, the heated cathode and the plate anode. This is a critical form of a nonlinear device. We still use semiconductor diodes for a variety of purposes.

2.2.2 Vacuum tube triode

Lee De Forest took the next step by inventing the **vacuum tube triode**, receiving US Patent 879,532 in 1908 for an amplifier based on the triode. As shown in Fig. 2.3, his tube added a grid between the cathode and anode. (He added the grid to avoid infringing on existing patents.) The voltage on the grid determined how much of the current generated by the cathode could reach the anode. The De Forest triode allowed us to control one electrical signal (the current from the cathode to the anode) with another (the grid). This was a critical step that allowed us to build amplifiers that could take a small signal and reproduce it on a larger scale.

Fig. 2.4 shows an idealized form of the characteristics of a vacuum tube. The two inputs to the tube are the plate voltage V_p between the cathode and anode and the grid voltage V_g from the grid to ground. The output variable is the plate current I_p from the cathode to the anode. For a given grid voltage, we can vary the plate current by changing the grid voltage. As we move to higher plate voltages, we increase the amount of plate current available. The graph shows a family of curves: each value of V_g gives a different I_p/V_p curve. In the plot, the slope of the I_p/V_g line is known as the **transconductance** g_m. We will see that modern MOS transistors have similar operating curves.

Tube amplifiers

Fig. 2.5 shows the schematic for a vacuum tube amplifier. The cathode and anode are connected across the power supply to form the output circuit. The voltage from the cathode to the anode forms the amplifier's output. We also need a resistor, known as the **load resistor**, along that path to transform the grid current to a voltage. The input signal is applied to the gate, with the voltage from gate to ground forming the input to be amplified. As we vary the grid voltage, the plate current varies under the relationship shown in Fig. 2.5. That plate current then causes variations in the resistor's voltage. Since the tube voltage is

$$V_p = V_{CC} - I_P/R_{L,}$$

the tube's output voltage varies in step with the plate current, which in turn depends on the grid voltage.

Radio used amplifiers to build **linear amplifiers** that could accurately reproduce a small signal, allowing it to be played back at a higher volume. Early computer designers used very similar circuits to build **nonlinear amplifiers** to manipulate digital signals.

Limitations

The vacuum tube was used prolifically in the well into the latter part of the 20th century. However, they did have several limitations that eventually led to the creation

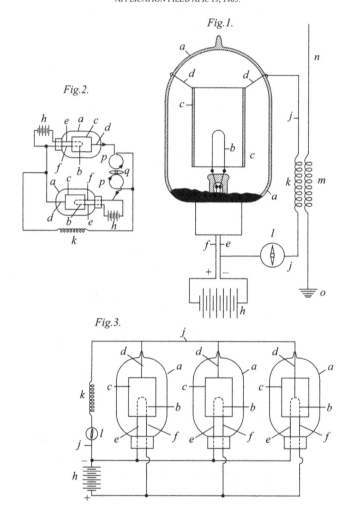

No. 803,684.

PATENTED NOV. 7, 1905.

J. A. FLEMING.
INSTRUMENT FOR CONVERTING ALTERNATING ELECTRIC CURRENTS
INTO CONTINUOUS CURRENTS.
APPLICATION FILED APR. 19, 1905.

FIGURE 2.2

The Fleming valve.

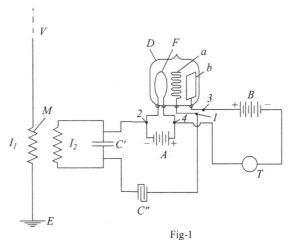

No. 879,532.   PATENTED FEB. 18, 1908.
L. DE FOREST.
SPACE TELEGRAPHY.
APPLICATION FILED JAN, 29, 1907.

Fig-1

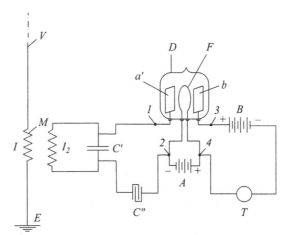

Fig-2

WITNESSES-
E. B. Tomlinson.
Patrick J. Conroy

INVENTOR:
Lee de Forest
by Geo. K. Woodworth,
Atty.

FIGURE 2.3

The De Forest triode.

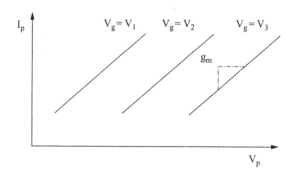

FIGURE 2.4

Idealized characteristics of a vacuum tube triode.

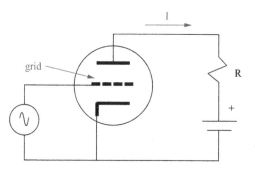

FIGURE 2.5

A vacuum tube amplifier.

of semiconductor devices. The physical size of tubes made systems built from them bulky. Even small tubes were the size of a thumb, and high-power devices could be much larger. Vacuum tubes also required relatively large voltages and consumed large amounts of power. But perhaps their greatest limitation came from their fragility. While the glass tubes could be broken, their filaments were their greatest source of failures. The filaments were physically small, many the width of a thread. When heated, they became particularly sensitive. Vibration or physical force could cause them to shatter. Even when they did not fail prematurely due to environmental stress, they eventually burned out through use. Electronic circuits had proved their usefulness, but the devices that made them possible needed to be more reliable.

2.3 Physics of materials

Before we look at semiconductor devices themselves, we need to develop a few concepts in physics. We will start with a model for conduction in metals. We will then develop the notion of temperature in Section 2.3.2; this topic will introduce Boltzmann's constant. Section 2.3.3 will introduce some models for the behavior of semiconducting materials.

Example 2.1 Physical Constants

Constant	Symbol	Value
Boltzmann's constant	k	1.38×10^{-23} J/K
Charge of an electron	q	1.6×10^{-19} C
Thermal voltage at 300K	kT/q	0.026 V
Permittivity of free space	ϵ_0	8.854×10^{-14} F/cm
Permittivity of silicon	ϵ_{Si}	$11.68\epsilon_0 = 1.03 \times 10^{-12}$ F/cm
Permittivity of SiO$_2$	ϵ_{ox}	$3.9\epsilon_0 = 3.45 \times 10^{-13}$ F/cm
Concentration of carriers in intrinsic silicon	n_i	1.45×10^{10} C/cm^3
Silicon effective density of states	N_c, N_v	$N_c = 3.2 \times 10^{19}$ cm^{-3} $N_v = 1.8 \times 10^{19}$ cm^{-3}
Silicon bandgap at 300K	E_g	1.12 eV

2.3.1 Metals

When atoms are close together, as in a solid, the nuclei of the surrounding atoms influence each atom's electrons. The result is that electrons occupy **bands**. Each band is a given range of energies; the farther away from the nucleus, the higher the energy of the electron. As shown in Fig. 2.6, we refer to the outmost band occupied when the atom is at minimum energy—absolute zero—as the **valence** band. The **conduction band** is above the valence band. Electrons with enough energy to occupy the conduction band are not tightly bound to one atom and can move from atom to atom.

As illustrated in Fig. 2.7, an electron is deflected from atom to atom as it moves. This form of motion is known as a **random walk** since the exact details of the movement of a particular electron are hard to predict—small variations in speed or angle can accumulate to large changes in the trajectory over time. Random walks can be observed macroscopically as Brownian motion of particles in water.

The **mobility** μ of the electron is defined as the ratio of the average time between collisions τ and the mass of the electron m_e [Fey10A]:

$$\mu = \frac{\tau}{m_e}. \tag{2.1}$$

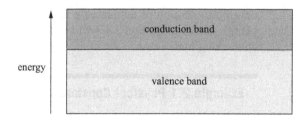

FIGURE 2.6

Energy bands in a conductor.

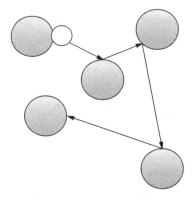

FIGURE 2.7

A random walk through a material.

When we apply an electric field to the metal, the field exerts a force on the electron and influences their random walk. As shown in Fig. 2.8, the electric field continually exerts a force on the electron, causing its trajectory to change somewhat at every step. At each collision, the particle acquires a new trajectory and velocity but is still under the effect of the electric field. We call the net motion of the electron under the influence of a field **drift**. If the time between collisions is τ and the mass of an electron is m_e, then the **drift velocity** is

$$v_{drift} = \varepsilon \frac{\tau}{m_e}. \tag{2.2}$$

The force on the electron is caused by the electric field ε caused by the voltage applied across the material:

$$\varepsilon = \frac{qV}{l} \tag{2.3}$$

We can write the drift velocity for the carriers in terms of mobility as

$$v_{drift} = \mu\varepsilon = \mu q \frac{V}{l}. \tag{2.4}$$

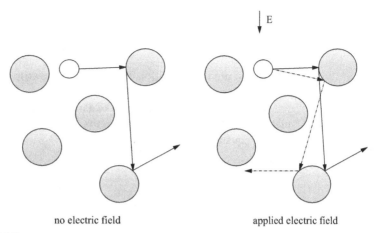

no electric field   applied electric field

FIGURE 2.8

A random walk influenced by an applied field.

Now consider the situation in Fig. 2.9. A block of has a length l and two faces with area A. The block is under the influence of an electric field from one face to the other. The electric field strength is V/l; we will assume that the field strength is constant across the length of the block. The average motion of a carrier is down the voltage gradient from one plate to the other. The current through the metal is charge per unit time:

$$I = qn_iAv_{drift} = \mu q^2 n_i \frac{A}{l} V \tag{2.5}$$

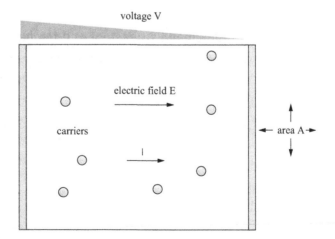

FIGURE 2.9

A model for drift current.

where n_i is the number of carriers per unit volume. We define resistance as

$$\frac{1}{R} = \mu q^2 n_i \frac{A}{l} \qquad (2.6)$$

We can use this definition of resistance to rewrite Eq. (2.5) in the traditional form of Ohm's law: $I = V/R$.

We often refer to the **resistivity** of a material to abstract away the shape of a particular sample. Resistivity is given in units Ω m. We can define resistivity in terms of the current equation:

$$\rho = \frac{1}{\mu q^2 n_i} \qquad (2.7)$$

We can define **conductivity** $\sigma = 1/\rho$.

Resistance depends on not only the properties of the material but also its shape. As shown in Fig. 2.10, current flows along the length of a piece of material. A longer piece of material gives the carriers more chance to interact with the material. A shape with a larger cross section gives more room for carriers to flow. We can write a relationship between resistivity of a material and the resistance of a particular shape of that material:

$$R = \rho \frac{l}{A}. \qquad (2.8)$$

Highlight 2.1

$R = \rho \frac{l}{A}$

2.3.2 Boltzmann's constant and temperature

Much of the underlying theory for the behavior of electrons comes from the theory of gases [Fey10A]. One of the fundamental physical constants we will deal with is

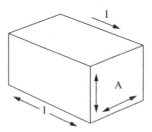

FIGURE 2.10

Resistance of a block of material.

Boltzmann's constant, known as k (or sometimes written as k_T); it relates temperature to energy.

A basic model for the energy of electrons comes from the behavior of gasses. The pressure in a cylinder of gas varies with height—the lower end of the cylinder is under greater pressure. The density of gas molecules n as a function of height h is

$$\frac{dn}{dh} = \frac{mg}{kT}n \qquad (2.9)$$

Solving this differential equation gives particle density as an exponential function of height:

$$n(h) = n_0 e^{-mgh/kT} \qquad (2.10)$$

The exponential form of this distribution is fundamental and will appear over and over in our equations.

Highlight 2.2

Many physical phenomena have the form $e^{energy/kT}$.

The **ideal gas law** relates the pressure P, volume V, and temperature T of a gas:

$$PV = NkT \qquad (2.11)$$

In this formula, N is the number of molecules in the gas. Boltzmann's constant is defined as

$$k = \frac{R}{N_A} \qquad (2.12)$$

where N_A is Avogadro's number ($N_A = 6.02 \times 10^{23}$ the number of molecules per mole) and R is energy per temperature unit per mole.

Temperature is a measure of the kinetic energy of molecules. We can understand the characteristics of temperature using the system shown in Fig. 2.11. Two chambers are separated by a moveable piston. Each temperature contains a gas; the two gasses need not be identical in composition, so their molecules may have different masses. However, if the average kinetic energies of the molecules in the two chambers are equal, we say they have the same temperature. Because the force exerted on each side of the piston is the same, it does not move. Absolute temperature is defined using Boltzmann's constant:

$$\text{mean molecular kinetic energy} = \frac{3}{2}kT \qquad (2.13)$$

The term 3/2 comes from a 1/2 contribution from each of the three degrees of freedom in the molecules' motion.

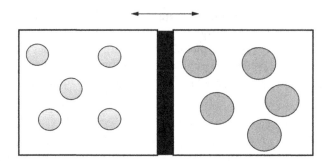

FIGURE 2.11

Two gasses separated by a movable piston as an example of temperature.

Boltzmann's constant plays a key role in physical phenomena throughout this book:

- It describes the number of electrons available at a given energy level, which determines properties like current.
- Energy distributions of electrons determine not just the current available for active computations but also leakage currents that waste energy and generate heat.
- It provides a link to the properties of materials as a function of temperature. We will see that heat generation and transmission play key roles in both small-scale and large-scale computing systems.

2.3.3 **Semiconductors**

Transistors can be understood using **solid-state physics**—the study of the behavior of matter in a solid state. Although many useful properties of gasses can be analyzed without resorting to quantum mechanics, the properties of semiconductors can be understood only through quantum mechanical concepts.

Silicon is one of a family of **semiconductors** (germanium is another) that provide us with important properties. Unlike metals, in which the valence and conduction bands overlap, the conduction band of a semiconductor does not overlap with the valence band. The situation is shown in Fig. 2.12. This **bandgap** means that electrons in a semiconductor need more energy to become conductive than is necessary in a metal.

At absolute zero (0K), the conduction band is empty; as the material is heated, electrons start to move to the conduction band. For an electron to jump from the valence to conduction band, it must have at least as much energy as the **gap energy** E_g, or the difference in energy at the bottom of the conduction band versus the top of the valence band. Silicon's bandgap is 1.12 electron volts (eV); 1 eV is the energy of one electron at a potential difference of 1 V. The thermal energy $kT = 0.026$ eV at room temperature, about 1/40 the value of the bandgap. We use 300K as an easy-to-manipulate value for room temperature; at that point, a significant number of electrons have enough energy to be promoted to the conduction and be available to conduct current. However, the greater energy required for an electron to be promoted

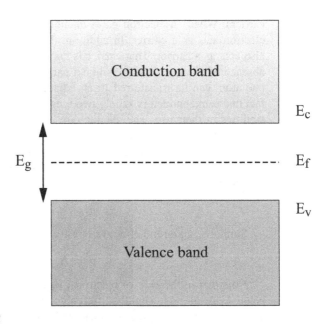

FIGURE 2.12

Conduction and valence bands in a semiconductor.

to the conduction band means that not as many electrons are available for conduction as in a metal, giving semiconductors higher resistivity.

Fermi level

One of the basic parameters of the band structure is the **Fermi level** E_f, which is the median energy level for electrons in the material. As shown in Fig. 2.12, the Fermi level for intrinsic silicon at 300K is between the conduction and valence bands; electrons do not exist there, but the probability distribution has its 0.5 level there. The population of electrons—the number of electrons at a given energy level—is governed by the Fermi-Dirac statistics [Tau98]:

$$f(E) = \frac{1}{1 - e^{(E - E_f)/kT}}$$
(2.14)

This formula is a modification of the Maxwell–Boltzmann statistics that underlie the analysis of temperature; its underlying assumptions are not valid in the case of electrons. The Pauli exclusion principle dictates that no two electrons can occupy the same state at the same time.

We often speak interchangeably about energy levels and potential levels. The electrostatic potential is related to energy levels by charge:

$$\psi = \frac{E}{q}$$
(2.15)

Holes

The bandgap of a semiconductor does much more than make the material less conductive. In fact, the gap allows us to control and manipulate the properties of carriers. When an electron gains enough energy to enter the conduction band, that electron acts as a carrier. In addition, the vacancy in the atom's electron structure also acts as a carrier. That carrier is known as a **hole**. Although it is created by the absence of an electron, it acts like a particle with a positive charge. The vacancy in one atom can be transferred to an adjacent atom, allowing the hole to move. The fact that semiconductors supply two types of conducting particles opens up new possibilities in their use. The conductivity of a semiconductor, while not as high as a metal, is high enough to make practical use of holes and electrons.

A natural semiconductor has an equal number of holes and electrons. If N_c is the effective state density of energy levels in the conduction band, the number of electrons is [Tau98]:

$$n = N_c e^{-(E_c - E_f)/kT} \qquad (2.16)$$

Similarly, given the effective density of states in the valence band,

$$p = N_v e^{-(E_v - E_f)/kT} \qquad (2.17)$$

Note that in the case of electrons, E_c is above E_f while the valence energy E_v is below E_f, resulting in slightly different forms for the formulas. We refer to pure silicon as intrinsic silicon (for reasons that will become clear shorty) and refer to the intrinsic carrier concentration as n_i.

Doping

We refer to undoped silicon as **intrinsic** material. Intrinsic semiconductors offer a limited range of options precisely because the number of holes and electrons are balanced. But we can change the proportion of the two carriers in a material by **doping**—adding impurities to the material. Semiconductors like silicon are in column IV of the periodic table; they have four electrons in their valence band. Arsenic is an example of a column V material with five electrons in the valence band. When silicon is doped with one of these **donor** materials, an extra electron is added without a corresponding hole, creating an *n*-type material. Column V materials are known as donors since they donate an additional electron. In comparison, boron is an example of a column III material with only three electrons in the valence band. When some of the silicon atoms are replaced with boron, the **acceptor** dopant creates holes without matching electrons, creating a *p*-type material. Column III materials are known as acceptor impurities. Adding dopants to crystalline silicon is not easy; techniques such as heating to extremely high temperatures or shooting the dopants in using an ion implantation gun are used to add them. But they give us extremely fine control over the properties of the material: they can be added to small regions; and the concentration of excess carriers is determined by the amount of dopant added to the silicon. Doped silicon is known as **extrinsic silicon**.

Carrier concentrations

Dopants affect the band structure and the distributions of carriers in the material. We can describe the carrier concentrations in terms of the doping levels. If N_d is the concentration of donor atoms and N_a is the concentration of acceptor atoms, then we can write the concentration of holes and electrons as

$$n = N_d e^{-(E_d - E_f)/kT} \tag{2.18}$$

$$p = N_a e^{-(E_f - E_a)/kT} \tag{2.19}$$

Doping moves the Fermi level: an n-type material has a higher Fermi level while p-type material has a lower Fermi level. We refer to the Fermi potential of intrinsic silicon as ψ_i. Rather than refer directly to the Fermi potential of a doped material, we often use ψ_B for the difference between the doped Fermi level and ψ_i:

$$\psi_B = |\psi_f - \psi_i| = \frac{kT}{q} \ln \frac{N_a}{n_i} \tag{2.20}$$

$$= \frac{kT}{q} \ln \frac{N_d}{n_i} \tag{2.21}$$

These two forms cover the cases of acceptor and donor doping, respectively.

We sometimes calculate the 0.5 occupancy energies separately for holes and electrons; those values are referred to as **quasi-Fermi levels** or **imrefs** (Fermi spelled backward). The imrefs are described by [Sze81]:

$$\phi_n = \psi - \frac{kT}{q} \ln \left(\frac{n}{n_i} \right) \tag{2.22}$$

$$\phi_p = \psi - \frac{kT}{q} \ln \left(\frac{p}{p_i} \right) \tag{2.23}$$

where ψ is the intrinsic level. At equilibrium, the quasi-Fermi levels are the same throughout the material. But when the material is in a nonequilibrium condition, such as an applied voltage, the quasi-Fermi levels may be different at various points in the material.

The product np is constant when the material is in equilibrium:

$$np = n_i^2 = N_c N_v e^{-(E_c - E_v)/kT} = N_c N_v e^{-E_g/kT} \tag{2.24}$$

This formula depends only on the gap energy, which is not affected by doping. We can use this relation to find the concentrations of carriers in doped materials. If N_d is the concentration of donor atoms in an n-type material, then

$$p = \frac{n_i^2}{N_d} \tag{2.25}$$

Similarly, for an acceptor concentration N_a in a p-type material,

$$n = \frac{n_i^2}{N_a} \tag{2.26}$$

Current   Current in semiconductors can come from a combination of two mechanisms, **diffusion** and **drift**. The total current is also the sum of the n and p currents, each of which has a diffusion and drift component.

The analysis of drift velocity at low applied fields is similar to the analysis of current in metals from Section 2.3.1. However, the determination of mobility in semiconductors is in general more complex than that for metals due to a variety of quantum mechanical effects. Holes and electrons have different mobility values with holes having lower mobility.

The diffusion current comes from variations in the concentration of particles. As shown in Fig. 2.13, if we have an open region with the right-hand side starting with more particles than the left-hand side, the motion of the particles will create a net flow from the high-density side to the low-density side. Even if each particle is equally likely to move in every direction, molecules on the high-density side are more likely to bounce to the low-density side than for the opposite to happen. The net current density (current per unit area of cross section) depends on the difference in the number of holes and electrons and their average velocity:

$$J_{diff} = q(n - p)v \tag{2.27}$$

In the x dimension, the net current is [Tau98]

$$J_{n,diff} = qD_n \frac{dn}{dx} \tag{2.28}$$

$$J_{p,diff} = qD_p \frac{dp}{dx} \tag{2.29}$$

The diffusion coefficient for carriers is related to their mobility through the Einstein relationship [Sze81]:

$$D_n = \mu_n \frac{kT}{q} \tag{2.30}$$

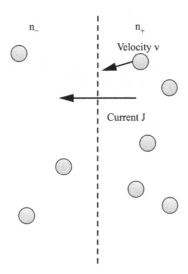

FIGURE 2.13

Diffusion current.

$$D_p = \mu_p \frac{kT}{q} \tag{2.31}$$

Drift and diffusion are linked in the case of electrons because electrons are charged particles.

2.4 Solid-state devices

The invention of the **transistor** provided the component that would drive a revolution in electronics. A group of scientists at Bell Laboratories decided to look into semiconductors as an alternative to vacuum tubes [Smi85]. The result of this effort was twofold. First, the field of solid-state physics was created to provide the foundation for understanding of the behavior of semiconductors; William Shockley made key contributions to this theory, including an early identification of the field effect as a possible mechanism for amplification. Second, the transistor was invented to solve the problems of the vacuum tube triode. John Bardeen and Walter Brattain invented the first practical device, the point-contact transistor [Bar50], and they demonstrated its use in an oscillator circuit on Christmas Eve, 1947. That first transistor is shown in Fig. 2.14. The three shared the Nobel Prize in Physics in 1956 for their work [Nob56].

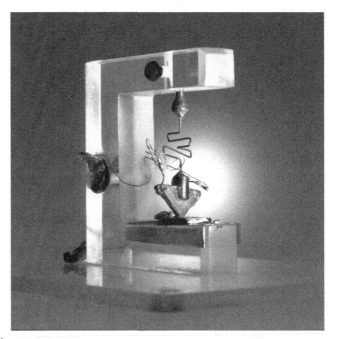

FIGURE 2.14

The first transistor.

Courtesy Alcatel-Lucent.

Modern computers use a different device, the MOSFET, but it still relies on the basic physical principles developed by Bardeen, Brattain, and Shockley.

This section will look at three types of devices: the semiconductor diode, the MOS capacitor, and the MOSFET.

2.4.1 Semiconductor diode

The **semiconductor diode** also has its own uses but its analysis gives some useful concepts for the MOSFET. It is a cousin of its vacuum tube predecessors, the Edison Effect tube and the Fleming valve.

As shown in Fig. 2.15, a semiconductor diode is built by putting together two differently doped pieces of silicon, one n-type and one p-type [Sze81]. The boundary between the n and p regions is known as the **junction** or **p-n junction**. The properties of this junction create the diode action of conducing current in one direction only.

Fig. 2.16 shows the structure of the bands at the p-n junction. The Fermi level is constant throughout the material, but the conduction and valence bands bend in response to the different amounts of positive and negative charge on either side of the region. On the p-type side, the conduction and valence band boundaries move up because there are fewer electrons at the higher levels; on the n-type side, the bands bend down thanks to the larger number of electrons in the conduction band. The figure

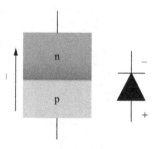

FIGURE 2.15

Structure and schematic symbol for a semiconductor diode.

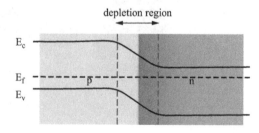

FIGURE 2.16

Band structures at a p-n junction.

also shows a **depletion region** around the junction. Near the boundaries, each type of carrier diffuses to the opposite region. The diffusion potential is known as the **built-in potential** [Sze81]:

$$V_{bi} = \frac{kT}{q}\ln\frac{N_a N_d}{n_i^2}. \tag{2.32}$$

As we apply a voltage across the junction, we change the levels of the conduction and valence bands. A positive voltage from p to n (**forward bias**) adds more holes to the p-type region, which then move toward the n-type region and creating a current through the diode. A negative voltage from p to n (**reverse bias**) sweeps holes out of the p-type region; this produces only a very small current in the reverse direction.

As shown in Fig. 2.17, applying an external bias voltage to the diode moves the band structures. A forward bias moves the n-type conduction band closer to the p-type's conduction band, allowing more current to flow. A reverse bias moves the n-type conduction band away from the p-type region's conduction band, increasing the energy barrier for carriers and reducing the current. In this nonequilibrium condition, the quasi-Fermi levels in the p and n materials are not the same; the depletion region serves as a transitional region between the two.

The combined n and p current densities $J = J_n + J_p$ includes both the drift and diffusion components. We saw the form of the drift component in EQ (2−5) hist-ohm-current-1 (normalizing for area to obtain the current density from current) and the diffusion currents in Eqs. (2.28) and (2.29). We can write the electron and hole current densities as the sum of these two components [Tau98]:

$$J_n = qn\mu_n\varepsilon + qD_n\frac{dn}{dx} \tag{2.33}$$

$$J_p = qn\mu_p\varepsilon + qD_p\frac{dp}{dx} \tag{2.34}$$

The current density Eqs. (2.33) and (2.34) give us one set of conditions that determine the current. We need to know the carrier concentrations to determine the diffusion current. The applied voltage is [Sze81]

$$V = \phi_p - \phi_n \tag{2.35}$$

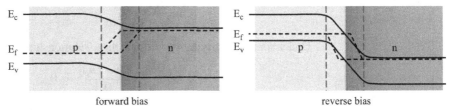

forward bias   reverse bias

FIGURE 2.17

Effect of applying bias voltage to a diode.

The imref relations allow us to write the concentrations of minority carriers at the edges of the depletion region (n concentrations in the p-type region, p concentrations in the n-type region):

$$n_p = n_{p0}e^{qV/kT} \tag{2.36}$$

$$p_n = p_{n0}e^{qV/kT} \tag{2.37}$$

where n_{p0} and p_{n0} are the equilibrium carrier concentrations. These relations form boundary conditions for the carrier distribution in the depletion region.

Some of the holes and electrons recombine. Since one hole recombines with one electron, we know that the recombination rates for the two are equal, allowing us to write a relationship between the n and p currents.

Conservation principles give us continuity equations for the minority carriers. For the case of electrons on the p-type side, the rate of change of electron concentration as a function of time depends on the current density gradient over space, the electron recombination rate R_n, and the electron generation rate G_n. The continuity equations for electron and hole minority carriers can be written as [Tau98]:

$$\frac{dn}{dt} = \frac{1}{q}\frac{dJ_n}{dx} - R_n + G_n \tag{2.38}$$

$$\frac{dp}{dt} = \frac{1}{q}\frac{dJ_p}{dx} - R_p + G_p \tag{2.39}$$

We can define the **lifetimes** of the carriers as

$$\tau_n = \frac{n - n_0}{R_n - G_n} \tag{2.40}$$

$$\tau_p = \frac{p - p_0}{R_p - G_p} \tag{2.41}$$

where n_0 and p_0 are the carrier concentrations at thermal equilibrium.

We can substitute the current density Eqs. (2.33) and (2.34) into the continuity equations:

$$\frac{dn}{dt} = n\mu_n\frac{d\varepsilon}{dx} + \mu_n\varepsilon\frac{dn}{dx} + D_n\frac{d^2n}{dx^2} - \frac{n_p - n_{p0}}{\tau_n} \tag{2.42}$$

$$\frac{dp}{dt} = p\mu_p\frac{d\varepsilon}{dx} + \mu_p\varepsilon\frac{dp}{dx} + D_n\frac{d^2p}{dx^2} - \frac{p_n - p_{n0}}{\tau_p} \tag{2.43}$$

Under some simplifying conditions, these relationships can be written as [Sze81]

$$\frac{d^2n_p}{dx^2} - \frac{n_p - n_{p0}}{D_n\tau_n} = 0 \tag{2.44}$$

$$\frac{d^2p_n}{dx^2} - \frac{p_n - p_{n0}}{D_p\tau_p} = 0 \tag{2.45}$$

When combined with the boundary conditions given by Eq. (2.36) and (2.37), we have the **Shockley equation** for the diode current density:

$$J = J_0 \left(e^{qV/KT} - 1 \right) \qquad (2.46)$$

$$J_0 = \frac{qD_n n_{p0}}{L_n} + \frac{qD_p p_{n0}}{L_p} \qquad (2.47)$$

where $L_n = \sqrt{D_n \tau_n}$ and $L_p = \sqrt{D_p \tau_p}$. The Shockley equation covers both the forward and reverse bias regions.

Fig. 2.18 shows the current versus voltage curve for the diode; this curve is often called an **IV curve**. At forward bias, the current increases exponentially. At reverse bias, the exponential term goes to zero and the reverse current asymptotically approaches a small negative value.

2.4.2 MOS capacitor

The **MOS capacitor** has other uses in itself, as we will see in Chapter 5, but is also a building block for the MOS transistor and is an example of an **electrostatic** device. The acronym **MOS** comes from **metal-oxide semiconductor**, referring to the conductor—insulator—conductor sandwich used to build a capacitor. The term MOS is applied to this structure even when the plate is formed by a material other than metal; for example, polycrystalline silicon (polysilicon) is often used to build the MOS capacitor structure.

As shown in Fig. 2.19, the structure is much like that of a classical capacitor, with an insulator separating two parallel plates [Tau98]. We refer to the bottom plate as the **substrate**, since that will be formed by the underlying silicon; the top plate, whether it be metal or silicon, is built on top of the silicon dioxide (SiO_2) insulator. Silicon dioxide is glass, the same basic material used in windows. As we apply a voltage across the two plates, we build up charge on one plate or the other, depending on the polarity of the voltage.

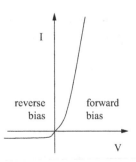

FIGURE 2.18

Current versus voltage characteristics of the diode.

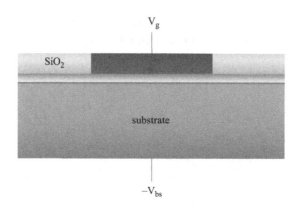

FIGURE 2.19

Structure of the MOS capacitor.

We will use the variable C_{ox} for capacitance per unit area and C for the capacitance of a pair of plates for a given area. The capacitance of the parallel plates of the MOS capacitor is computed in the same way as for a classical capacitor:

$$C_{ox} = \frac{\epsilon_{ox}}{t_{ox}} \qquad (2.48)$$

where ϵ_{ox} is the permittivity of silicon dioxide and t_{ox} is the thickness of the oxide. The parallel plate capacitance is inversely proportional to the thickness of the oxide. We will see that this capacitance has mixed implications for logic gates: increased load but also increased transistor current.

Highlight 2.3

MOS oxide capacitance is inversely proportional to oxide thickness.

Example 2.2 MOS Capacitor Values

We will assume that our capacitor is of size $L = 32$ nm, $W = 64$ nm; nonsquare devices are common in modern manufacturing.

If the oxide thickness is 1.65 nm, then taking into account the area of the capacitor plates,

$$C = WL\frac{\epsilon_{ox}}{t_{ox}} = \left(32 \times 10^{-7} \text{ cm}\right)\left(64 \times 10^{-7} \text{ cm}\right)\left(\frac{3.45 \times 10^{-13} \text{ F/cm}}{1.65 \times 10^{-7} \text{ cm}}\right)$$

$$= 4.3 \times 10^{-17} \text{ F} = .043 \text{ fF.}$$

Example 2.3 MOS Capacitor Technology Trends

Here are some typical values for parameters of the MOS parameter in several different technologies [PTM15]:

Technology (nm)	t_{ox} (nm)	C_{ox} (F/cm²)
130	2.25	1.53×10^{-6}
90	2.05	1.68×10^{-6}
65	1.85	1.86×10^{-6}
45	1.75	1.97×10^{-6}
32	1.65	2.09×10^{-6}
20	1.4	2.46×10^{-6}
16	1.35	2.56×10^{-6}
10	1.2	2.88×10^{-6}
7	1.15	3.00×10^{-6}

The MOS capacitor's behavior is more complex than that of the classical capacitor because due to the presence of both holes and electrons in the substrate. Fig. 2.20 shows what happens as we apply a voltage to the upper plate in a p-type material. (If we used an n-type material for the substrate, the voltage polarity would have to be reversed and the roles of holes and electrons would switch, but the same basic phenomenon would still occur.) At zero applied voltage, which we refer to as **flatband**, the electrons and holes are at their normal concentrations all the way up to the boundary of the insulator. As we apply a positive voltage, the conduction and valence bands

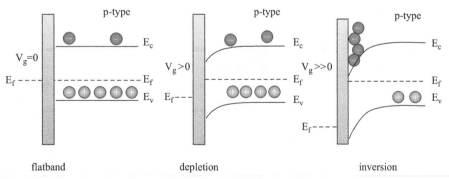

FIGURE 2.20

Operation of the MOS capacitor [Sze81].

bend relative to the Fermi level, reflecting the fact that electrons are being driven away from the plate's boundary and holes are being drawn toward it. We refer to this moderate change in carrier concentrations at moderate voltages as **depletion**—the majority n-type carriers have been depleted near the plate's boundary. As we further increase the voltage, the bands bend further. At some point, p-type carriers, which were in the minority in the unbiased material, now have the same concentration as the electrons did at zero voltage. We refer to this condition as **inversion** since the populations of the two carrier types have become inverted. The inversion layer is extremely thin, less than 50 Å. Note that far away from the plate, the carriers go back to their normal distributions. But at the plate boundary, we have used an applied voltage to control which type of carrier is available for conducting. We will make use of the ability to select between n-type and p-type carriers in the MOS transistor.

Changes in the charge distribution cause changes in the potential $\psi(x)$ as a function of distance x from the surface. The potential describes the band bending as a function of position. We call $\psi(0) = \psi_s$ the **surface potential**. The potential in the bulk is $\psi(\infty) = 0$. In the case of electrons in a p-type material, the density of the minority carriers as a function of depth is

$$n(x) = \frac{n_i^2}{N_a} e^{q\psi(x)/kT} \tag{2.49}$$

This equation implies that carrier concentration is an exponential function of voltage.

The **threshold potential** is the potential at which the carrier population is **inverted**. We can quantify the meaning of inversion in any of several ways, but the standard criterion for inversion is that the minority carrier population is equal to the doping concentration: $n_{inv}(0) = N_a$ for our example of electrons as minority carriers. This criterion means that the minority carrier concentration at the surface is equal to the depletion charge density. When we substitute $n_{inv}(0) = N_a$ into Eq. (2.49), we obtain this expression to find the threshold potential for our example of electrons in p-type material:

$$\psi_{s,inv} = 2\psi_B = 2\frac{kT}{q}\ln\frac{N_a}{n_i} \tag{2.50}$$

Highlight 2.4

Minority carrier density under inversion in the MOS capacitor is an exponential function of the surface potential.

Defects in the oxide hurt the operation of the capacitor. Imperfections at the boundary between the silicon and silicon dioxide layers trap carriers, reducing the number of carriers available for inversion. Charge trapped in the oxide must be overcome to achieve inversion, increasing the threshold voltage. The key challenge in the invention of the MOS transistor was the manufacture of very pure oxides.

We now have a structure that we can use to control the carrier concentrations in one part of the device using a voltage from another signal.

2.4.3 Basic MOSFET operation

Structure

The type of transistor that now dominates computer design is the **MOS field-effect transistor** or **MOSFET**. The term field-effect transistor refers to the way in which the device operates, using a voltage to create an electric field to control currents. The first MOSFET was created by Dahwon Kahng and Martin Atallah at Bell Labs in 1959 [Ata60; Kah63; Kah76].

Figs. 2.21 and 2.22 show side and top views of the transistor, respectively. The side view shows that the transistor is a combination of the other two structures we studied: the MOS capacitor and the diode. The region under the capacitor oxide is known as

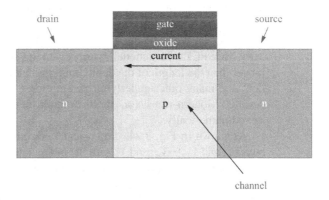

FIGURE 2.21

Cross section of an *n*-type MOS transistor.

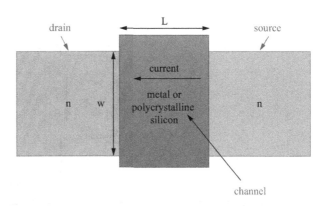

FIGURE 2.22

View from above of an *n*-type MOS transistor.

the **channel**; this is where the transistor action happens. The regions at the two ends of the channel are known as the **source** and **drain**. We can actually build two types of transistors: if the source and drain are n-type, we have an **n-type transistor**, named after the minority carrier; if the source and drain are doped p-type, the transistor is **p-type**. (The transistor type more specifically refers to the type of the minority carrier in the channel which is the same as the doping of the source and drain.) This diagram also shows the bands across the transistor. The top view shows two important physical measurements of the transistor, its **length** and **width**. Both are measured relative to the direction of current flow. Many of the other physical parameters of the transistor, such as oxide thickness, are set by the manufacturing process. Transistor length and width, however, can be chosen by the circuit designer to optimize the circuit properties; they are the knobs we will use to optimize logic delay and energy consumption. Fig. 2.23 shows the schematic symbol of both n-type and p-type transistors. It also shows the major voltages and currents of interest: I_d is the drain-to-source current; V_{ds} is the drain-to-source voltage; V_{gs} is the gate-to-substrate voltage.

Device model

A simple model of the IV characteristics of the MOSFET is sufficient for most of our purposes. We will concentrate the *long-channel* model which was developed for early MOSFETs. This model assumes, among other things, that the channel is much longer than the length of the depletion region around the p-n junctions. That assumption, and many other underlying assumptions, does not hold for modern nanometer devices. Modern transistors require much more complex models that can only be solved numerically.

As shown in Fig. 2.24, the current I_d through the transistor channel depends on both V_{ds} and V_{gs}—V_{ds} and V_{gs} are the independent variables while I_d is the dependent variable. Fig. 2.26 shows the plot of drain current I_d versus drain-to-source voltage V_{ds} with and gate-to-substrate voltage V_{gs} as an additional variable. Below the **threshold voltage** V_t, we assume that no drain current flows. (We will revisit this simplification shortly). For a given V_{gs}, the current first grows as the drain-to-source voltage is

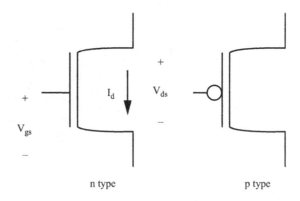

FIGURE 2.23

Symbol for the transistor and its associated voltages and currents.

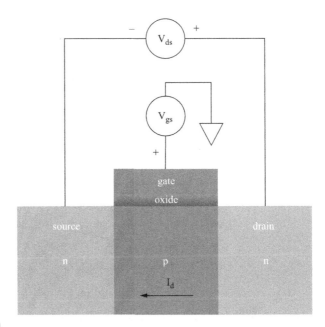

FIGURE 2.24

Voltages and currents in the MOSFET.

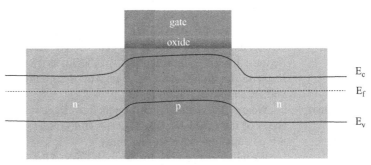

FIGURE 2.25

Energy bands in the *n*-type MOSFET.

increased; we call this the **linear region** even though, as we will see, the current is not quite linear. After V_{ds} reaches a certain level, the device enters the **saturation region**, the drain current flattens out. V_{gs} determines the level of the saturation current and the slope of the linear region current growth.

Fig. 2.25 shows the configuration of bands in the *n*-type MOSFET with no applied voltages. The higher band levels in the channel form an energy barrier that prevents electrons from the drain moving into the channel. V_{ds} and V_{gs} move the bands. A positive V_{gs} lowers the bands near the surface in the channel; a positive V_{ds} lowers the

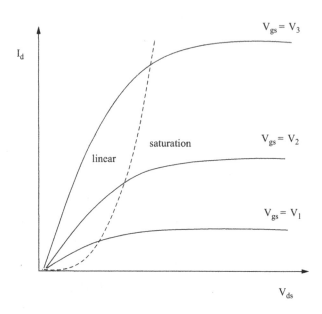

FIGURE 2.26

Current versus voltage curves for the MOS transistor.

bands in the drain. Both these conditions lower the energy barrier and help electrons to flow through the device.

The form of the MOSFET's IV curve is shown in Fig. 2.26. We need a family of curves, one for each value of the gate voltage. Each of those curves is divided into two regions: **linear** and **saturation**. In the linear region, the drain current is (roughly) linear in the drain/source voltage. The drain current is independent of drain/source voltage in the saturation region.

Device model derivation   We can derive the transistor equations to gain some insight into the physical properties of the transistor [Sze81]. We will concentrate on the n-type device here; the p-type equations have similar form.

We will use μ_n and μ_p for the mobilities of n-type and p-type carriers; keep in mind that the mobility of carriers in the channel is considerably lower than in bulk silicon. For example, the drift mobility at 300K in silicon is $\mu_n = 1.05 \times 10^3 \text{ cm}^2/\text{V s}$ [Sze81, p. 29]. The inversion layer mobility decreases with increasing drain-to-source voltage; it varies from a high of 800 cm^2/V s at a field of 1×10^5 V/cm^2 to a low of 400 cm^2/V s at a field of 6×10^5 V/cm^2 at a temperature of 25°C [Sze81 p. 449].

This simple model of the MOSFET is based on Ohmic resistance with the added twist that the carrier concentration can be modulated by the applied gate voltage. The conductivity of the transistor's channel as a function of depth x is related to the charge density and mobility:

$$\sigma(x) = qn(x)\mu_n(x). \tag{2.51}$$

If we assume that mobility is constant across the channel, then we can compute the channel conductance g from the conductivity by integrating over depth:

$$g = q\mu_n \frac{W}{L} \int_0^{x_i} n(x)dx = q\mu_n |Q_n| \frac{W}{L}. \tag{2.52}$$

where $|Q_n|$ is the total charge in the vertical slice of the channel and W and L are the width and length of the channel.

The incremental resistance of a horizontal section of the channel dy is:

$$dR = \frac{dy}{gL} = \frac{dy}{W\mu_n |Q_n(y)|}. \tag{2.53}$$

The voltage drop across that section is

$$dV = I_d dR = \frac{I_d dy}{W\mu_n |Q_n(y)|}. \tag{2.54}$$

To compute the current, we need to know the charge. When we analyzed the threshold voltage of the MOS capacitor, we were only concerned with surface charge since we were interested in the point at which that surface charge population inverted. To understand the current through the MOSFET, we must consider both the surface charge and the charge in the bulk silicon.

If we make several simplifying assumptions (no interface traps or fixed charge, pure drift current, etc.), then we can derive a relatively simple equation for the charge. ψ_s is the surface potential at the start of the strong inversion region of operation; it can be approximated by $\psi_s = V_D + 2\psi_B$.

The MOS capacitor that forms the gate draws charge toward the plate. Some of that charge contributes to the inversion layer while other charge is deeper into the bulk. The total charge is given the basic capacitor relation. We will refer to the gate capacitance as C_g to be consistent with common usage in circuit design but remember that $C_g = C_{ox}$. The charge in the inversion layer is the difference between the total charge and the bulk charge:

$$Q_n(y) = Q_s(y) - Q_b(y) = -[V_G - V(y) - 2\psi_B]C_g + \sqrt{2\epsilon_{si}qN_A[V(y) + 2\psi_B]}. \tag{2.55}$$

We use this approach for a more complete accounting of the channel charge. When we analyzed the threshold voltage of the MOS capacitor, we based that criterion solely on the charge at $x = 0$. In the case of the MOSFET, the channel current includes minority carriers that are not precisely at the interface. Eq. (2.55) takes into account all of the charge that contributes to drain current.

We can find a general form of the current by integrating Eq. (2.54) over voltages $[0, V_{ds}]$ and channel positions $[0, L]$:

$$W\mu_n \int_0^{V_{ds}} |Q(n)|dV = I_{ds} \int_0^L dy \tag{2.56}$$

$$W\mu_n \int_0^{V_{ds}} \left\{ [V_G - V(y) - 2\psi_B]C_g + \sqrt{2\epsilon_{si}qN_A[V(y) + 2\psi_B]} \right\} dV = I_{ds}L \quad (2.57)$$

We can write the current equation in this form because I_{ds} is constant across the length of the channel due to charge conservation. However, the applied drain-to-source voltage seen across the channel varies with distance: at the source, the gate voltage is 0; at the drain, $V = V_{ds}$. The result is

$$I_{ds} = \frac{W}{L}\mu_n C_g \left\{ \left(V_{gs} - 2\psi_B - \frac{1}{2}V_{ds} \right) V_{ds} \right.$$
$$\left. - \frac{2}{3}\frac{\sqrt{\epsilon_{si}qN_A/\psi_B}}{C_g} \left[(V_{ds} + 2\psi_B)^{3/2} - (2\psi_B)^{3/2} \right] \right\}. \quad (2.58)$$

This equation for charge holds for all voltages that cause strong inversion. We can simplify it by considering two cases: small applied drain-to-source voltage V_{ds} and large V_{ds}. Using the case of small V_{ds}, we can identify a formula for the threshold voltage V_t [Tau98]:

The threshold voltage has two components, the inversion charge component from Eq. (2.50) and a separate term for the bulk charge:

$$V_{tn} = 2\psi_B + \frac{\sqrt{2\epsilon_{si}qN_a(2\psi_B)}}{C_{ox}} = 2\frac{kT}{q}\ln\frac{N_a}{n_i} + \frac{\sqrt{2\epsilon_{si}qN_a(2\psi_B)}}{C_{ox}} \quad (2.59)$$

Note that the threshold voltage is inversely proportional to gate capacitance. Below the threshold voltage, we assume for this model the gate does not conduct. Our discussion of the MOS capacitor explained why the cutoff region exists—the channel carrier population has not yet been inverted.

When we substitute this back into Eq. (2.58) and simplify for small V_D, we find the linear region current:

$$I_d = k'\frac{W}{L} \left[(V_{gs} - V_t)V_{ds} - \frac{1}{2}V_{ds}^2 \right] \quad (2.60)$$

Fig. 2.27 helps us understand why the drain current saturates. The figure shows the transistor at three different values of V_{ds}; the gate voltage is constant and above the threshold voltage in all cases. At $V_{ds} = 0$, the inversion layer has the same thickness

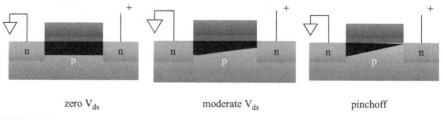

zero V_{ds} moderate V_{ds} pinchoff

FIGURE 2.27

Evolution of the minority carrier populations with increasing source-to-drain voltage.

throughout. As we raise V_{ds}, the inversion layer has thinned out at the terminal with the positive voltage. At the boundary between the linear and saturation regions, the inversion layer at the positive terminal has reduced itself to zero thickness. This condition is called **pinchoff**. The channel will continue to conduct a current, but the pinchoff condition limits the amount of current that we can draw.

We can find the saturation region current equation by finding the pinchoff voltage, which we do by substituting $Q_n(L) = 0$ into Eq. (2.55) and using the notation $K = \sqrt{\varepsilon_{si} q N_a}/C_g$:

$$V_{D,sat} = V_G - 2\psi_B + K^2 \left(1 - \sqrt{2V_G/K^2} \right).$$

(2.61)

When we substitute this back into the full drain current formula, we find the saturation current:

$$I_d = \frac{1}{2} k' \frac{W}{L} (V_{gs} - V_t)^2$$

(2.62)

By convention, the n-type drain current flows from drain to source.

The drain current does change slightly with increased drain-to-source current in saturation. This effect, known as body effect, is not significant for the circuit models we will develop in Chapter 3.

Device model summary   We use the notation $k'_n = \mu_n C_{ox}$ and $k'_p = \mu_p C_{ox}$ as the **transconductance** of the device. The term *transconductance* comes from the fact that it relates the output current to an input voltage; typical units are $\mu A/V^2$ or A/V^2. Note that increasing the gate capacitance increases the transistor's transconductance, which in turn increases the amount of current it produces. The drain current is proportional to the transistor's width and inversely proportional to its length. We sometimes use the notation

$$\beta = k' \frac{W}{L}.$$

(2.63)

The equations for the p-type transistor have the same form but opposite signs on most of the values: drain current flows from source to drain, the threshold voltage is negative, and we refer to V_{sd} and V_{sg}. We can create negative threshold voltages without adding an additional power supply by connecting the substrates of the p-type and n-type devices to opposite polarities. As shown in Fig. 2.28, the n-type

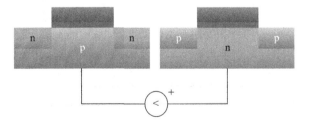

FIGURE 2.28

Biasing of n-type and p-type transistors.

transistors are built in a p substrate that is connected to the negative power supply terminal; the p-type transistors are in an n-type substrate connected to the positive terminal. When a gate voltage within the normal power supply bounds is applied to each of these transistors, the result is a gate voltage of the proper polarity: positive for the n-type and negative for the p-type.

We can now write the drain current equations for the three different regions of operation:

cutoff $V_{gs} < V_t$	$I_d = 0$	(2.64)
linear $V_{ds} < V_{gs} - V_t$	$I_d = k'\frac{W}{L}\left[(V_{gs} - V_t)V_{ds} - \frac{1}{2}V_{ds}^2\right]$	(2.65)
saturation $V_{ds} \geq V_{gs} - V_t$	$I_d = \frac{1}{2}k'\frac{W}{L}(V_{gs} - V_t)^2$	(2.66)

Highlight 2.5

MOSFET transconductance is proportional to gate capacitance.

Example 2.4 MOS Transistor Currents

We can compute some sample values based on our calculations for the MOS capacitor; we will assume our transistor's channel is of the same size, $L = 180$ nm, $W = 270$ nm.

The transconductance of our n-type transistor is

$$k_n' = 170 \ \mu A\Big/V^2$$

We computed the threshold voltage of our MOS capacitor to be $V_t = 0.7$ V. If we connect a voltage source of $V_{ds} = 0.3$ V across the drain and source and connect the gate to $V_{gs} = 1.1$ V, then the current at this point in the linear region is

$$I_d = \left(170 \ \mu A/V^2\right)\frac{3}{2}\left[(1.1 \ V - 0.7 \ V)0.3 \ V - \frac{1}{2}0.3 \ V^2\right] = 19 \ \mu A$$

If $V_{gs} = 1.2$ V and $V_{ds} = 1.2$ V, then the transistor is in the saturation region and

$$I_d = \frac{1}{2}\cdot\left(170 \ \mu A/V^2\right)\cdot\frac{3}{2}\cdot(1.2 \ V - 0.7 \ V)^2 = 32 \ \mu A.$$

Example 2.5 MOS Transistor Parameter Trends

Here are typical transistor parameter values for both n-type and p-type transistors for a range of technologies [PTM15]:

Technology (nm)	V_{tn} (V)	V_{tp} (V)	μ_n (cm²/ V s)	μ_p (cm²/ V s)	k'_n (A/V²)	k'_p (A/V²)
130	0.38	−0.32	0.059	0.0084	9.1×10^{-4}	1.2×10^{-4}
90	0.40	−0.34	0.055	0.0071	9.2×10^{-4}	1.2×10^{-4}
65	0.42	−0.37	0.049	0.0057	9.2×10^{-4}	1.1×10^{-4}
45	0.47	−0.41	0.0440	0.0044	8.7×10^{-4}	0.87×10^{-4}
32	0.51	−0.45	0.0389	0.0036	8.1×10^{-4}	0.74×10^{-4}
22	0.51	−0.37	0.0181	0.0023	5.2×10^{-4}	0.66×10^{-4}

Note that n-type and p-type transistors have different parameters even in a single technology generation. The threshold voltages for p-type transistors are generally lower. The transconductance of p-type transistors is also lower due to the lower effective mobility of holes relative to electrons. We can also see that the parameter values for p-type and n-type diverge as transistors shrink.

2.4.4 Advanced MOSFET characteristics

We are interested in several properties of the MOSFET that are not directly considered by the long-channel model: leakage and temperature sensitivity. We will also consider some alternatives to the planar MOSFET that have been designed to meet some of the challenges of nanometer-scale MOSFETs.

Leakage

One major category of static power consumption has become increasingly important as transistor geometries have shrunk. **Leakage current** is any current that is not controlled by the transistor gate. A number of physical mechanisms cause leakage current. As devices shrink, leakage currents become a larger proportion of total currents. In many chips made with modern manufacturing processes, total leakage current is larger than the total dynamic power consumption of the logic.

Subthreshold current

Our assumption of cutoff is a simplification. Drain current does not disappear immediately when the gate voltage drops below the threshold. Since there are still minority carriers in the channel even below threshold voltage, some drain current can be carried. This **subthreshold current** is an increasingly important source of leakage current.

As shown in Fig. 2.29, the drain current drops off exponentially below V_t. We characterize the shape of this curve using **subthreshold slope**, which we typically write using the **subthreshold swing** [Sze81; Tau98]:

$$S = 2.3 \frac{kT}{q} \left(1 + \frac{C_{dm}}{C_{ox}} \right)$$

$$(2.67)$$

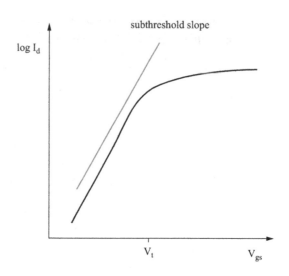

FIGURE 2.29

Subthreshold current characteristics.

In this formula, C_{dm} is the depletion layer capacitance. Remember that the inversion charge depends exponentially on V_{gs}, which explains the semilog behavior of subthreshold current. The subthreshold slope is independent of V_{ds}.

We want a small value for S, which corresponds to a sharp dropoff in subthreshold current. Increasing C_{ox} will have the desired effect on subthreshold current but will also increase the capacitive load presented by the gate, increasing delay.

The drain current has two components, drift and diffusion, with diffusion current dominating in the subthreshold region. The diffusion current depends on charge density:

$$I_d = \mu_{eff}\frac{W}{L}\sqrt{\frac{\epsilon_{Si}qN_a}{2\psi_s}}\left(\frac{kT}{q}\right)^2\left(\frac{n_i}{N_a}\right)^2 e^{q\psi_s/kT}\left(1 - e^{-qV_{ds}/kT}\right) \qquad (2.68)$$

This current is insensitive to device parameters, such as device size or doping. Subthreshold current does not scale. Its magnitude is insensitive to device sizes. As transistors have become smaller, their saturation currents have decreased while subthreshold currents remain the same. What was once a trivial phenomenon is now a major problem in computer design. Subthreshold currents will become a larger and larger part of total current. Leakage current has grown exponentially and is now larger than total dynamic current.

DIBL

Subthreshold current occurs even at zero source/drain voltage; drain-induced barrier lowering (**DIBL**) further increases leakage when the source/drain voltage is increased. Our assumption that transistors turn completely off below the threshold voltage is optimistic, particularly for modern devices. Early MOS devices could be modeled as **long-channel devices** in which the channel length was long compared to any effects of the source and drain regions. As Moore's Law has progressed, the source and drain have come to show more effect on the operation of the channel. The source and drain are *p-n* junctions, which have depletion regions just as we saw in the diode. When the channel is short enough, those depletion regions start to affect the current in the channel, with their electric fields lowering the potential barrier in the middle of the channel. A source/drain voltage causes further lowering of the potential barrier, increasing the subthreshold drain current.

Temperature   Given the form of the device equations, it should be no surprise that transistor operation is affected by temperature. The threshold voltage of a MOSFET decreases with increasing temperature [Sze81], which we can see by differentiating the threshold voltage formula:

$$\frac{dV_t}{dT} = \frac{d\psi_B}{dT}\left(2 + \frac{1}{C_g}\sqrt{\frac{\epsilon_{si}qN_a}{\psi_B}}\right) \tag{2.69}$$

$$\frac{d\psi_B}{dT} = \pm\frac{1}{T}\left[\frac{E_g(T=0)}{2q} - |\psi_B(T)|\right] \tag{2.70}$$

Subthreshold slope decreases with temperature. The leakage current of a MOS-FET at 100°C is 30–50× larger than its leakage current at 25°C [Tau98]. Both these facts are bad for chip operation. As a chip heats up, drain currents and subthreshold currents will both increase. That increased power consumption will result in higher temperatures. In the extreme case, the result is **thermal runaway** that can cause permanent damage to the chip. The subthreshold current, which is always flowing even when the logic gates are not changing, causes the chip to dissipate heat. That heat then increases the subthreshold current, causing more heat. The result can be heat so intense that the chip destroys itself.

Highlight 2.6

Leakage current increases with temperature.

Alternative structures   Leakage currents have become so large that manufacturers have radically changed the transistor structure to reduce leakage. Fig. 2.30 shows the structure of a finFET. This transistor is not a planar structure. Instead, the transistor is fabricated on a fin of silicon that extends above the surface of the wafer. The gate wraps around the fin to create a channel region, which allows the gate voltage to control the carriers in the narrow thin much better than is possible in a bulk transistor. An alternative is the silicon-on-insulator (SOI) structure shown in Fig. 2.31. Early SOI technologies grew a thin layer of silicon on the top of a substrate of a different type, such as sapphire, but alternative substrates provide to be too costly. Modern SOI structures grow an oxide layer upon which the transistor is built. The thin channel region helps the

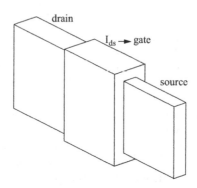

FIGURE 2.30

Structure of a finFET.

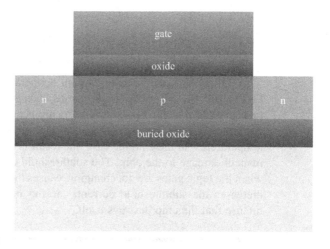

FIGURE 2.31

Cross section of a silicon-on-insulator (SOI) transistor.

gate to control electrons. Reduced drain-induced barrier lowering is one benefit of SOI structures.

2.5 **Integrated circuits**

The next, critical step to building modern computers was the development of the **integrated circuit** or **IC**. The transistor's importance was hugely magnified by its role in the IC; transistors on printed circuit boards would still have improved on tubes, but electronic systems would still be bulky and power-hungry. Today, thanks to the huge advantages of integrated circuits, more transistors are manufactured each year in the state of California than raindrops fall on the state.

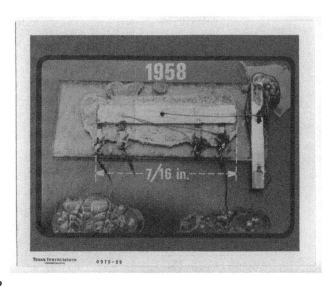

FIGURE 2.32

The first integrated circuit, courtesy DeGolyer Library, Southern Methodist University, Texas Instruments records.

Various aspects of the integrated circuit were invented by Jack Kilby [Kil64] and Robert Noyce [Noy61]. Kilby won the Nobel Prize in Physics in 2000 for his work [Nob00B]. Kilby's first integrated circuit is shown in Fig. 2.32.

The original thinking that led to the IC was startling in its directness. Transistors were being fabricated in batches on a common substrate, which was then chopped into small pieces, one per transistor. The transistors were then assembled and wired together on printed circuit boards. Why cut apart the transistors? Why not put the wiring on top of the substrate to create a single, integrated circuit with devices and wires?

We will first discuss Moore's Law, which describes the exponential rate in the reduction in device geometries. We will then survey IC manufacturing processes in Section 2.5.2 followed by a more detailed discussion of lithography in Section 2.5.3. The next section discusses the mathematics of yield. Finally, Section 2.5.5 discusses the separation of concerns between manufacturing and circuit design.

2.5.1 Moore's Law

The first and most obvious limit to how many transistors we can put on a single chip is the size with which we can make features on the wafer. Gordon Moore, cofounder of Intel, understood early in the history of the integrated circuit that feature sizes on chips were shrinking steadily, which meant that the number of transistors per chip were growing. That result is now known as **Moore's Law**, which has held for over a half century. As shown in Fig. 2.33, the rate of change has varied a little over the

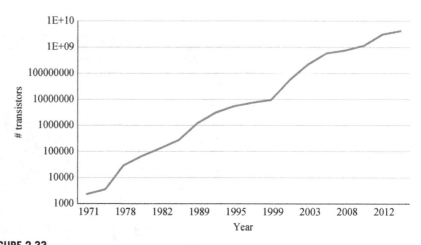

FIGURE 2.33

Moore's Law [Tra15].

years, but for many years the number of transistors per chip has doubled every 18 months. Because the growth in transistor count compounds, the number of transistors per chip grows exponentially.

A manufacturing process requires the development of an ensemble of related technologies. A given manufacturing process is known as a **technology node**. The dimension that defines the node is traditionally determined by the length of the smallest transistor that can be manufactured by that node. However, as manufacturing has grown more complex, the relationship between feature sizes and node names has become more complex.

Fig. 2.34 shows how transistor manufacturing has improved over the years. The data come from the reports of the International Technology Roadmap for Semiconductors (ITRS), an industry planning group. The plot shows the industry goals for

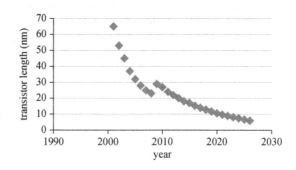

FIGURE 2.34

Transistor length over time [ITR01; ITR05; ITR07; ITR11; ITR13].

fabricated length of transistors in high-performance microprocessors over a period of 20 years. Transistor sizes have shrunk at an exponential rate.

Moore's Law ensures that chips will become more complex over time. But that does not necessarily mean that they will be faster. As we will see in the next chapter, another analysis predicted that chips would become faster as the transistors shrank. This combination of more and faster transistors has been the huge economic motivation that has kept Moore's Law in force for over 50 years.

2.5.2 Manufacturing processes

Modern IC fabrication lines process large wafers, like the one shown in Fig. 2.35. Each wafer contains many chips, as can be seen on the surface of this wafer. The wafer is then cut apart, the chips are tested, and good chips are put into electronic packages that provide protection and mechanically stable wiring. The chips are then ready to be used in systems.

IC fabrication is known as **planar processing** because much of the structure lies in a series of parallel planes. The planar process was developed by Jean Hoerni of Fairchild Camera and Instrument for bipolar transistors [Hoe62], but the basic concepts are equally applicable to MOS transistors. The transistor channels are created using the wafer as the substrate, with dopants added to control its electrical characteristics. A series of layers of different materials are put on the top to create the wires and the transistor gates; silicon dioxide is used as insulation between those layers. Fig. 2.36 shows a simple chip from top and side views. The chip includes a transistor, a metal wire, and a **via** that connects the channel's drain to the metal wire. The cross section of the transistor is as we expect. The via consists of a vertical hole in the glass that is

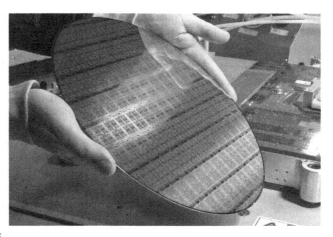

FIGURE 2.35

A wafer covered with integrated circuits.

Courtesy of International Business Machines Corporation, © International Business Machines Corporation.

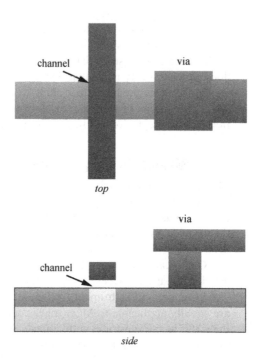

FIGURE 2.36

A chip layout from top and side views.

filled with metal; the metal flows down to the substrate to connect to the drain and fills up to the level of the metal wire at the other side of the connection.

The placement of material on each layer is specified by a **mask** that is used to make patterns on the wafer. Modern chips use 7−10 layers of wiring stacked vertically.

Chips are built in facilities known as **clean rooms** for their careful control of dust and other contaminants. Given the tiny sizes of the devices being made, even the smallest amount of material can cause manufacturing errors. Modern VLSI fabrication lines cost several billions of dollars to build, most of that cost coming from the very precise equipment required. That huge investment is possible because of the huge volumes of chips that can be made on a fab line.

Fig. 2.37 shows the main flow of the manufacturing process. Most processing steps consist of three major steps. First, a pattern is put on the wafer using a **mask** that describes the features to be put on during this step. The pattern controls where material will be deposited. That material is then processed in some way: heat treatment for dopant diffusion, for example. The major steps in the fabrication include preparing the wafer, adding diffusions to the substrate, applying polysilcion wires, forming the metal wires, and finally passivating the chip with a layer of silicon dioxide to protect it against the elements.

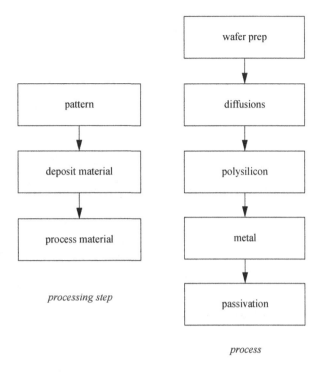

FIGURE 2.37

The fabrication process.

Given the extremely small dimensions and huge numbers of components involved, it should not be surprising that not all chips work. A variety of test procedures are used to identify working chips. **Functional testing** checks the Boolean behavior of the circuits while **parametric testing** tests properties such as clock frequency and operating temperatures.

Once the chip is finished, it is put into a **package** that protects it against the elements and handling. The package provides physical support as well as a set of **pins** to connect to the printed circuit board. A set of **pads** on the chip provide regions of metal large enough to provide connections to the package. Good electrical connections require solid mechanical properties that are provided by the pins to the chip on one side and the printed circuit board on the other. Packages are made from either plastic or ceramic, depending on the environmental and temperature specifications. Fig. 2.38 shows an Intel Pentium Pro from 1995. This processor consisted of two chips: the CPU is on the left and the cache is on the right. The chips are held in a cavity; in the standard case, this cavity would be covered with a lid to protect the chips. Rows of pads can be seen around the edges of the chips that provide the connection point for wires that connect to the chips. Pins extend vertically to connect to the board. This package is made of ceramic; in the case of a plastic package, a pad frame

FIGURE 2.38

Intel Pentium Pro package c.1995.

connects the chip to the pins, then plastic is molded around the chip and pad frame. Fig. 2.39 shows an Intel Broadwell processor from 2014. This package also includes two chips, but the newer design is both smaller and lighter. The chips in the Broadwell package are protected by lids and are not visible in the photo.

FIGURE 2.39

Intel Broadwell package c.2014.

Courtesy Intel Corporation.

2.5.3 **Lithography**

Lithography refers to the creation of patterns on the surface of the wafer. The properties of light and optical systems form one of the fundamental limitations on IC manufacturing.

Masks are projected onto the wafer using a lens system. The wafer is coated with a photosensitive material that captures the image. The resolution of the masking process is determined primarily by the optics. The wave nature of photons causes the pattern of irradiance projected by the lens to be affected by refraction [Hec98]. As shown in Fig. 2.40, the width of the aperture that controls access to the lens is d and the distance to the chip surface is D. The wavelength of the light is λ. The irradiance I as a function of the angle θ away from the centerline is complex but has the form

$$I(\theta) \propto \text{sinc}^2\theta \tag{2.71}$$

The form of the diffraction pattern formed by the lens is the basis for Rayleigh's criterion, a basic criterion for resolution. As shown in Fig. 2.41, Rayleigh's criterion states that two features projected by the optical system are resolvable if the center of the diffraction pattern of one feature is aligned with edge of the central maximum

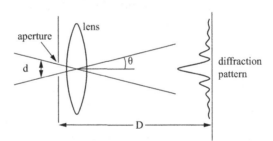

FIGURE 2.40

A lens projecting a mask onto a wafer.

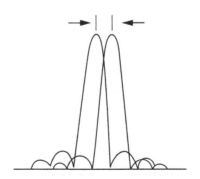

FIGURE 2.41

Rayleigh's criterion.

bright spot of the second feature's pattern. The diameter of the central maximum bright spot is $1.22\lambda D/d$. The ratio d/D is known as the numerical aperture NA, so we can rewrite this formula as

$$\Re = 1.22\frac{\lambda}{NA}. \tag{2.72}$$

For optical systems of the type commonly used in photolithography, the Rayleigh factor is commonly written in the form

$$\Re = k_1\frac{\lambda}{NA} \tag{2.73}$$

where k_1 summarizes several physical details of the optical system and is about 0.75 for typical lithographic setups.

The wavelength of the illuminating light provides a fundamental limit on the resolution of the optical system. Remember that wavelength and frequency of light are related by $f = c/\lambda$. Higher frequencies of light, those toward the ultraviolet end of the spectrum, provide higher resolution.

Example 2.6 Lithographic Resolution

A krypton fluoride (KrF) laser has a wavelength of 248 nm. A typical value for the numerical aperture for a mask projection system is 0.6. This gives us a minimum resolution of

$$\Re = 0.75\frac{248 \times 10^{-9}}{0.6} = 310 \text{ nm}$$

This simple analysis of resolution suggests that IC manufacturers should have moved to extreme ultraviolet light sources to achieve the resolutions required for modern feature sizes. But using these light sources is expensive, so manufacturers have developed a number of techniques to extend the useful light of older lithographic technologies.

Resolution depends upon numerical aperture. The required numerical aperture is determined by the refractive index of the medium between the lens and the target. While reducing the size of the aperture increases resolution, it also increases the exposure time required to print the mask since less light passes through the aperture. Longer exposure times increase the time required to print the mask; given that modern processes may use over 100 masking steps, increasing exposure time could substantially increase manufacturing time. We can take advantage of the dependence of NA on the refractive index of the medium in front of the lens by changing that medium. By using fluid rather than air, we can increase the numerical aperture and reduce the minimum resolvable feature. Immersion optics have allowed manufacturers to substantially reduce the size of the minimum feature that can be fabricated without changing the wavelength of the light used to illuminate the masks.

Computational lithography analyzes the effects of diffraction and creates a mask whose projection will result in the desired pattern—the mask's pattern cancels out the unwanted effects of diffraction. **Double-patterning** creates fine features in two steps, one for each side of the feature.

2.5.4 Yield

Physical processes show natural variations which cause the problem of **yield** in manufacturing processes. Some devices may not operate properly, meet performance specifications, or show other problems. Given the small sizes of the transistors and wires in modern ICs, yield is an important problem in semiconductor manufacturing [Gup08]. Many different problems can cause a chip to fail: holes in the transistor gate oxide; over- or underdiffusion of the source and drain; wires that are too thin or too thick; wires that are too narrow or wide; vias that do not open up; etc.

A simple model for yield is the Poisson distribution. Given a density of defects per unit area D, a chip is good if no defects fall in its area A. The yield is

$$Y = e^{-AD} \tag{2.74}$$

However, this model tends to be pessimistic because many types of faults are not distributed evenly across the chip.

Defects can be created at any step in the manufacturing process, so the overall yield of the process depends on the yield of each step:

$$Y = \sum_{1 \leq i \leq n} Y_i \tag{2.75}$$

Given that modern manufacturing processes may have over 100 steps, the yield of each step must be very high to achieve an acceptable yield of finished chips.

In the most basic case, the failure of any single transistor, wire, or via is enough to render the chip useless. However, we can sometimes make use of chips with failures. In some cases, such as multicore processors, we can simply not use the failed components (and charge less for the chip). In other cases, such as memory, we add extra circuits during the original manufacturing process, test the chip, and perform minor alterations to substitute the spare circuits for the bad circuits if necessary.

Yield is influenced not only by functional characteristics such as shorted wires but also **parametric** failures such as speed or power. Circuits must be designed to work not just with devices at **nominal** parameter values but also across a range of device parameters. Manufacturing specifies ranges of parameters that they can deliver with reasonable assurance. These parameters define a multidimensional space; we often refer to the extreme values of that space as the **process corners**. Circuits must be designed to work at all those process corners—deliver adequate speed, power consumption within spec, etc. As we have seen, the device parameters are closely related—for example, both threshold voltage and transconductance depend upon gate capacitance. As a result, not all process corners are equally likely to occur.

Process variability

Process variability is another important determinant of yield. Manufacturing variations can cause significant variations in the parameters of interest for logic reliability, performance, and energy consumption.

Operating temperature is an example of an important parametric specification. While process designers do not have direct control over the operating temperature of a chip, they must keep in mind that device characteristics depend on temperature, often strongly. As a result, they must specify a temperature range over which the device will operate. Circuit designers must consider the effects of temperature on circuit operation.

Example 2.7 Sensitivity of $I_{d,sat}$

This plot shows variation of $I_{d,sat}$ for $\pm20\%$ variation of the threshold voltage and transconductance. These parameters vary around the nominal values $V_t = 0.5$ V, $k' = 8.1 \times 10^{-4}$ A/V:

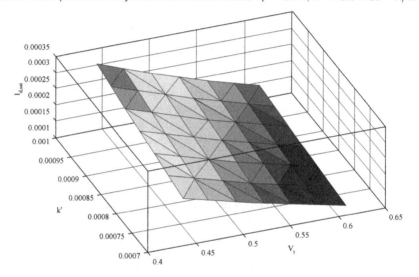

2.5.5 Separation of concerns

A manufacturing process is designed to support a range of digital designs; we typically do not create a process for a particular chip. (Dynamic RAM is a major exception—they require specialized devices and circuits.) This separation of concerns allows us to spread the huge cost of developing the process and building the manufacturing plant over several designs; it also helps to mitigate the risk of any particular design.

However, it also means that we do not have total freedom in our designs. The digital designer cannot easily ask the process designers, for example, to change the characteristics of the transistor. The process designers for their part cannot mandate that certain types of circuits be used or avoided.

Roughly speaking, manufacturing is responsible for the characteristics of the transistors and wires; circuit designers create circuits using those components. The circuit designers, for example, cannot alter the transconductance of the transistors. They can, however, design the *W/L* of each transistor based on the needs of its circuit.

Highlight 2.7

Some of the important relationships between device parameters:

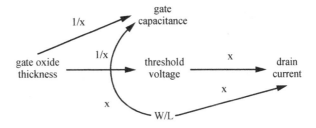

2.6 Synthesis

- MOS transistors are characterized by their threshold voltage and transconductance. Manufacturing process designers determine values for threshold voltage, etc., based on application requirements and physical limitations. Circuit designers can choose the transistor channel W/L to adjust its characteristics.
- Transistor parameters are related to each other. Changing one parameter also changes other transistor parameters.
- Integrated circuits combine transistors and wires to create complete circuits and systems.
- Moore's Law is an observation on the rate of improvement of integrated circuit technology. The size of transistors decreases exponentially and the number of transistors per chip increases exponentially.

Questions

Q2-1 The resistivity of copper is $1.7 \times 10^{-8}\ \Omega$ m and its carrier concentration is 8.5×10^{28} m^{-3}. What is its mobility?

Q2-2 You are given a copper wire with cross-sectional area $A = 6.25 \times 10^{-16}$ m^2 and length 5×10^{-7} m. The resistivity of copper is $1.6 \times 10^{-6}\ \Omega$ cm. What is the resistance of the wire?

Q2-3 Use Eqs. 2.18 and 2.19 to plot the difference between Fermi level and intrinsic Fermi level at 300K for a range of doping:
  a. $10^{15} \leq N_d \leq 10^{18}$ cm^{-3}.
  b. $10^{15} \leq N_a \leq 10^{18}$ cm^{-3}.

Q2-4 How many electrons are required to charge a capacitor to a voltage of 1 V in each case:
  a. Capacitance is 10 fF?
  b. Capacitance is 0.1 fF?

Q2-5 Plot C_{ox} as a function of 1 nm $\leq t_{ox} \leq$ 5 nm for $W = 45$ nm, $L = 30$ nm.

Q2-6 An MOS capacitor is doped at $N_a = 10^{15}$ cm^{-3}. What is its threshold potential at 300K?

$$\psi_s(inv) = 2\frac{kT}{q}\ln\frac{N_a}{n_i} = 2\frac{(1.38 \times 10^{-23})(300)}{1.6 \times 10^{-19}}\ln\frac{10^{15}}{1.45 \times 10^{16}} = -0.14 \text{ V}$$

Q2-7 You are given the conduction and valence bands and the Fermi level for several devices. In each case, draw the boundaries between doping regions and identify the type (n or p) of each region.

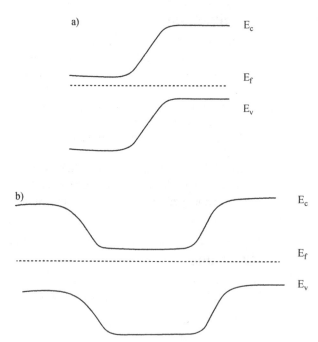

Q2-8 You are given a simplified form of the threshold voltage formula $V_t = P + \frac{Q}{C}$ with arbitrary constants P and Q.
  a. If we reduce the W and L sizes of the plates by ½ but do not change the oxide thickness, how does the threshold voltage scale?
  b. How does threshold voltage scale if you scale W, L, and oxide thickness x by the same amount?

Q2-9 You are given a transistor with $k' = 80 \ \mu A/V^2$, $V_t = 0.5 \ V$, $W/L = 1.5$. Plot the boundary between the transistor's linear and saturation regions for $0 \le V_{ds} \le 1 \ V$.

Q2-10 What is the drain current for each transistor for $|V_{gs}| = |V_{ds}| = 1.1 \ V$?
  a. n-type, $V_{tn} = 0.5 \ V$, $k'_n = 80 \ \mu A/V^2$, $W/L = 2$.
  b. p-type, $V_{tp} = -0.5 \ V$, $k'_p = 50 \ \mu A/V^2$, $W/L = 2$.

Q2-11 Compute the saturation drain current for n-type and p-type with $|V_{gs}| = |V_{ds}| = 1 \ V$, $W/L = 3$:

n-type:

a)	$V_{tn} = 0.55 \ V$, $k'_n = 75 \ \mu A/V^2$
b)	$V_{tn} = 0.45 \ V$, $k'_n = 85 \ \mu A/V^2$
c)	$V_{tn} = 0.50 \ V$, $k'_n = 75 \ \mu A/V^2$

p-type:

d)	$V_{tp} = -0.55 \ V$, $k'_p = 55 \ \mu A/V^2$
e)	$V_{tp} = -0.45 \ V$, $k'_n = 30 \ \mu A/V^2$
f)	$V_{tp} = -0.50 \ V$, $k'_n = 45 \ \mu A/V^2$

Q2-12 Compute the linear region drain current for n-type and p-type with $|V_{gs}| = 1 \ V$, $|V_{ds}| = 0.3 \ V$, $W/L = 3$:

n-type:

a)	$V_{tn} = 0.55 \ V$, $k'_n = 75 \ \mu A/V^2$
b)	$V_{tn} = 0.45 \ V$, $k'_n = 85 \ \mu A/V^2$
c)	$V_{tn} = 0.50 \ V$, $k'_n = 75 \ \mu A/V^2$

p-type:

d)	$V_{tp} = -0.55\ \text{V},\ k'_p = 55\ \mu\text{A}/\text{V}^2$
e)	$V_{tp} = -0.45\ \text{V},\ k'_n = 30\ \mu\text{A}/\text{V}^2$
f)	$V_{tp} = -0.50\ \text{V},\ k'_n = 45\ \mu\text{A}/\text{V}^2$

Q2-13 Plot drain current as a function of V_{ds} for various values of V_t. Assume an n-type transistor with $k' = 80\ \mu\text{A}/\text{V}^2,\ V_t = 0.5\ \text{V},\ W/L = 1$:

$$V_{gs} = 0.75\ \text{V}$$

$$V_{gs} = 1\ \text{V}$$

Q2-14 A leakage current discharges a capacitor that is initially charged to a voltage of 1 V.
  a. How much leakage current produces a 10% change in a 1 fF capacitor's voltage over 1 ps?
  b. How much leakage current for a 50% change for a 0.1 fF capacitor over 10 ps?

Q2-15 Find values related to Moore's law:
  a. How many 18-month intervals have occurred between 1971 and 2016?
  b. Write a function for transistor count as a function of year y with the number of transistors in 1971 as $n_{1971} = 3000$.
  c. What exponential weight gives 10 billion transistors in 2016?

Q2-16 Plot Rayleigh's criterion for minimum resolution over a range of wavelength $200 \leq \lambda \leq 400$ nm. Assume $k_1 = 0.75,\ NA = 0.6$.

Logic Gates

3

3.1 Introduction

In this chapter, we will study the fundamental physical properties of logic gates. We will do so by studying a single type of logic gate: the CMOS inverter. We have three goals. First, we want to understand why digital logic is designed the way it is. Second, we want to understand the design space for logic gates—how changing one of the design parameters affects the quality of the gate. Finally, we want to understand how Moore's Law affects the properties of gates from technology generation to generation.

We are interested in three physical properties of logic gates and of digital systems in general: **performance**, **energy**, and **reliability**. Logic gates have other physical properties. The area required to build the gate, for example, is an important factor in its cost. But area is primarily a topic for the design of particular gates; we are most interested in the logic design space.

Performance

Performance is a term that is used in many fields to mean many different things. In computing, performance usually means speed. In the case of logic gates, we are interested in the amount of time a gate takes to transform a change in its inputs to the new value at its output. The speed of logic gates does not directly translate into CPU performance or software performance. But logic performance does form a bound on the speed with which larger systems can operate. Hardware designers spend a great deal of time optimizing the performance of their designs.

Energy

Changing the state of a physical system takes *energy*. Logic gates are no different. When a logic gate evaluates its inputs and generates an output, it consumes energy. Energy has its own cost—more energy per gate operation means a larger electric bill. But energy consumption has many other effects. Power, or energy per unit time, determines the amount of heat generated on a chip. Excess heat dissipation can cause a variety of problems, chip failure among them. The huge amounts of power required by some chip also create engineering problems. Power and performance are at odds: higher performance means higher power.

Reliability

The importance of *reliability* may not be so obvious if only because, to the average user, computers seen to rarely work improperly. But digital logic does fail. Given that the transistors on a high-end chip are only a few hundred atoms long, the fact that they

The Physics of Computing. http://dx.doi.org/10.1016/B978-0-12-809381-8.00003-1
Copyright © 2017 Elsevier Inc. All rights reserved.

may fail should not come as a huge surprise. We can engineer digital systems to be highly reliable but we cannot engineer them to be free of failure. Fundamental physical limits govern the reliability of digital systems.

Section 3.2 introduces the CMOS inverter and Section 3.4 develops basic concepts about logic circuits. Section 3.3 examines the inverter's static properties. Section 3.4 models inverter delay while Section 3.5 models power. Section 3.6 introduces ideal scaling theory. Section 3.7 studies models for gate reliability.

3.2 The CMOS inverter

Complementary MOS or **CMOS** circuits dominate digital design today. The term *complementary* comes from the use of both p-type and n-type transistors in a single gate. CMOS came to prominence for two reasons: it was relatively easy to manufacture and it consumed little power. Modern CMOS requires much more complex manufacturing processes and, as we will see, power consumption is a major concern in modern CMOS chips. But the reign of CMOS is unlikely to end any time soon.

Fig. 3.1 shows the schematic for a static, complementary inverter as well as the symbol we use for an inverter. The gates of the two transistors are connected together. (Unfortunately, we use *gate* for both transistors and logical circuits. If the meaning is unclear, we will specify).

CMOS requires both n-type and p-type transistors, which means that different parts of the substrate require different doping. The most common way to build CMOS is **twin-tub** processing which, as shown in Fig. 3.2, masks off different parts of the substrate for implantation of the proper dopants.

The basic operation of the inverter is simple to understand. While both transistor gates are connected together, n-type and p-type respond to opposite polarities, so they

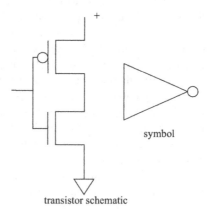

transistor schematic

FIGURE 3.1

Schematic for a CMOS inverter.

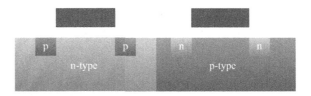

FIGURE 3.2

Cross section of twin-tub CMOS.

will perform opposite functions. If the input is a high voltage, the n-type transistor will be on and the p-type will be off; as a result, the output will be low. If the input is a low voltage, the n-type is off and the p-type is on, producing a high output voltage. The n-type is called the **pulldown** transistor and the p-type is known as the **pullup**. The two terminals of the power supply for CMOS circuits are called by tradition V_{DD} for the positive and V_{SS} for the negative. V_{SS} is often connected to ground, and we will generally refer to the power supply voltage simply as V_{DD}. In most cases, we will simplify our equations by assuming $V_{SS} = 0$.

We often use the term **ground** for a common connection that is given a zero voltage value. But that reference of zero voltage is not chosen arbitrarily. The earth provides an excellent voltage reference; **earthed grounds** are connected electrically to the earth itself. Earth is used as a voltage reference because it is very hard to change the voltage of the earth. If charge is placed at a point on the surface of a metal sphere, electrostatic forces cause the charge to be evenly distributed around the sphere, giving a very small contribution of charge at any point. The earth is not an ideal conductor. It is, however, an extremely large conductor that can absorb a huge amount of charge without noticeably changing its voltage.

Electrical waveforms are continuous functions of time. The Turing machine model requires discrete values, not continuous values like voltages. To use our logic gates to build Turing machines, we need to figure out how to represent discrete values using their continuous waveforms. As illustrated in Fig. 3.3, the logic gates will still operate on voltages and currents. But we will no longer be concerned about the exact value of a gate's output voltage. We will instead use a range of signal values to represent a single bit.

While some logic families use currents to represent logical values, CMOS lends itself to using voltages. Using a range of values to represent a bit makes our circuits resistant to noise, which can come from many different sources. Fig. 3.4 shows how we map the range of voltages between the two extremes of the power supply voltages: low voltages correspond to logic 0 and high voltages correspond to logic 1. We call the upper boundary of the logic 0 range V_L; the lower end of the logic 1 range is called V_H. We also have a range of voltages that correspond to neither 0 or 1. We call this range X. Assigning the middle voltages to neither 0 or 1 is a natural consequence of how the inverter works, as we will see shortly.

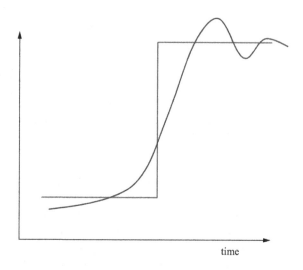

FIGURE 3.3

Waveforms and discrete values.

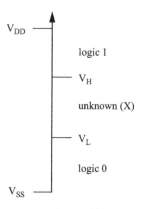

FIGURE 3.4

Voltages and logical values.

X versus don't-cares   The unknown value X is a fundamentally different concept from the don't-cares of Boolean logic. Boolean functions can make use of two different types of don't-cares. An *input don't-care* is a shorthand for a set of related minterms. An *output don't-care* is an incompletely specified function in which the function's output value for some input combination is chosen based on minimization opportunities; the fully realized function always gives the same output for that set of inputs. In contrast, an unknown value in a circuit is a voltage—a known, measurable value—that does not correspond to a logic 0 or 1.

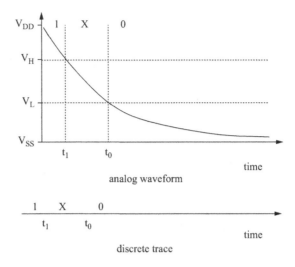

FIGURE 3.5

A waveform and its logical values.

Fig. 3.5 shows a simple waveform over time and how it corresponds to Boolean values. At time t_1, the waveform's voltage crosses from the logic 1 range to the X range. At time t_0, it crosses the boundary into the logic 0 range. If we watched the output on an instrument that displayed only digital values, we would see the signal start at 1, change to X at t_1, and then to 0 at t_0.

3.3 Static gate characteristics

Some of the important properties of the inverter come from its static characteristics—its operation when its inputs do not change. Logic levels are determined by static characteristics. Logic levels in turn affect some important reliability characteristics of the inverter.

Transfer curves   How do we choose the values for V_H and V_L? The standard approach is to use a circuit characteristic known as the **transfer curve**. A voltage transfer curve shows the relationship between input and output voltages; time is not involved. Since the transfer curve is not a function of time, we consider it a **static** property of the gate. We can measure the transfer curve of a circuit by applying a sequence of voltages and measuring the output for each after we have given the output time to settle to its final value. We choose several input voltages; then for each input voltage, we evaluate the output voltage. We need to write a series of equations that relates the currents and voltages of the two transistors.

Fig. 3.6 shows the transfer curve for an inverter. The curve displays the inverting characteristic: a low input voltage gives a high output voltage and *vice versa*.

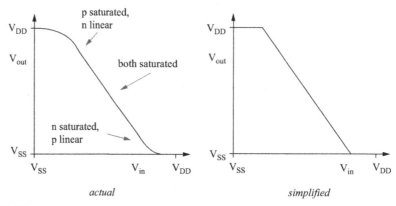

FIGURE 3.6

Voltage transfer curve for a CMOS inverter.

In between those two extremes, the transfer curve has a high slope. By comparing the voltages on the axes of the transfer curve to the inverter schematic, we can see how the transistors respond. At a low input voltage, the n-type transistor is in the linear range and the p-type is saturated. At a high input voltage, the n-type is saturated and the p-type is linear. Both are fully on and saturated only at a middle of the curve. In typical operation, the inverter spends most of its time at either end of the transfer curve.

We can find the transfer curve by computing it using our equations for the transistor drain currents. Because the curve has a complex shape, we will sometimes use a simplified version of the transfer characteristics. As the figure shows, we can simply approximate the transfer curve as a three piecewise linear segments: one for the high output region, one for the low output, and one for the transition.

Fig. 3.7 shows the transfer curves for two inverters. One inverter has n-type and p-type transistors of equal size; the other has transistors of equal β, which required the p-type transistor to be about four times larger than the n-type. The transconductances of the n-type transistors in the two inverters are equal, as are the p-type transconductances. Changing the relative sizes of the pullup and pulldown transistors does not substantially affect the slope of the transfer curve at its middle, which we use to characterize gain. It does, however, substantially move the transfer curve—making the pullup larger moves the transfer curve to the right.

The simplest way to create the transfer curve is to generate a set of points. The p-type transistor's IV characteristics have the same basic form as the n-type, but V_{gs} and V_t are both negative. (We can apply a negative voltage to the gate without going outside the limits of the power supply by connecting the p-type's substrate to V_{DD}. Since the substrate for the p-type is doped n and the n-type substrate is doped p, connecting each region to a different power supply terminal does not cause problems). We will label each variable with its transistor type, for example $V_{t,p}$.

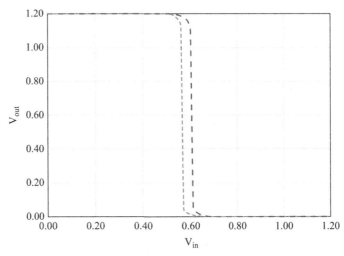

FIGURE 3.7

Transfer curves as a function of pullup size.

Finding logic levels

We use the transfer curve to determine the logic level voltages by finding the points on the curve at which its slope is −1. There are two such points, one at high output voltages and one at low output voltages. This procedure actually gives us separate values for input and output voltages. As shown in Fig. 3.8, V_{IH} is the minimum voltage for a logic 1 at the input while V_{OH} is the minimum logic 1 voltage at the output.

Noise margin

To ensure that gates operate properly, we need to be sure that the output levels of one gate are compatible with the input levels of the next. As shown in Fig. 3.9, we want to be sure that $V_{OH} \geq V_{IH}$ and $V_{OL} \leq V_{IL}$—we want to be sure that the output of one gate produces a valid input for the next gate. Note that the high and low logic levels need not be symmetric. The difference between the required input and output voltages is known as the **noise margin**. For example, the logic high noise margin is

$$NM_H = V_{OH} - V_{IH}. \tag{3.1}$$

The logic low noise margin is

$$NM_L = V_{IL} - V_{OL}. \tag{3.2}$$

Examination of the transfer curve shows that CMOS gates easily meet the noise margin requirement. As shown in Fig. 3.10, a voltage very just below V_{IL} will be far away from the power supply voltage, but the inverter converts that to an output voltage that is very close to the power supply. We often refer to this characteristic as **restoring** logic values. The gain of the logic gate is essential to the restoring properties of logic gates. Logic gates amplify input signals (perhaps including inverting their sense, but still amplifying in magnitude) and so push voltage values closer to

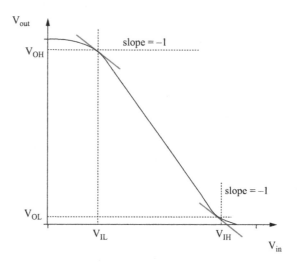

FIGURE 3.8

Choosing the voltages for logic levels.

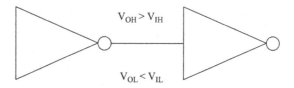

FIGURE 3.9

Compatibility of output and input voltage levels.

the power supply. When the input voltage is in the vicinity of one of the power supply values, then even large variations in input voltage will cause only small changes in the output. We refer to these types of amplifying logic gates as **saturating** logic because their voltages tend to saturate at the power supply terminals.

Larger noise margins mean more noise is required to turn a valid logic value into an invalid value. We can derive a simple relationship between transistor parameters and noise margins if we assume that the device transconductances are equal or $k'_n = k'_p$:

$$NM_L = \frac{3}{8}\left(V_{DD} + V_{t,p} - V_{t,n}\right) \tag{3.3}$$

This formula shows that increasing the threshold voltage improves noise margins:

$$NM \propto V_t \tag{3.4}$$

(The same proportionality holds if the transconductances are not equal, but the formula becomes messier). Of course, increasing the transistor thresholds also increases gate delay.

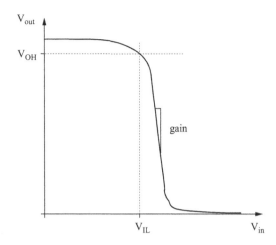

FIGURE 3.10

Gain and restoring logic values.

Middle voltage

We usually find the values for the logic levels graphically; finding an algebraic formula for the logic levels is messy. A simpler formula that gives us some insight into the operation of the inverter is the **middle voltage** V_M. As shown in Fig. 3.11, this voltage is at the midway point of the transfer characteristic. We can solve for V_M by writing down some equations for the inverter. The gates of the two transistors are connected:

$$V_{gs,n} = V_{DD} + V_{gs,p}. \tag{3.5}$$

The currents through the two transistor channels are also equal:

$$I_{d,n} = I_{d,p}. \tag{3.6}$$

The sum of the voltages across the two channels equals the power supply voltage:

$$V_{DD} = V_{ds,p} + V_{ds,n}. \tag{3.7}$$

At the middle of the transfer characteristic, both of the transistors are in the saturation region, so we can expand Eq. (3.6) using the definitions of the saturation drain current:

$$\frac{1}{2}\beta_n\left(V_{gs,n} - V_{tn}\right)^2 = \frac{1}{2}\beta_p\left(V_{DD} - V_{gs,p} - |V_{tp}|\right)^2. \tag{3.8}$$

By convention, we say that the n-type's source is connected to ground and the p-type's source is connected to V_{DD}. In a simple MOSFET, the source and drain are symmetric; in more advanced transistor designs, the source and drain have somewhat different structures.

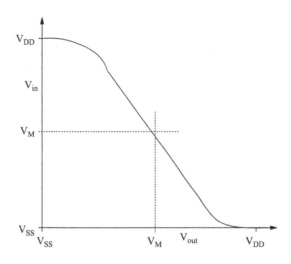

FIGURE 3.11

The middle voltage V_M of the transfer characteristic.

We can substitute in V_M for the gate voltage:

$$\frac{1}{2}\beta_n(V_M - V_{tn})^2 = \frac{1}{2}\beta_p(V_{DD} - V_M - |V_{tp}|)^2. \tag{3.9}$$

We can now solve for V_M:

$$V_M = \frac{\sqrt{\frac{\beta_p}{\beta_n}}(V_{DD} - |V_{tp}|) + V_{tn}}{1 + \sqrt{\frac{\beta_p}{\beta_n}}}. \tag{3.10}$$

The value of V_M is a function of the transconductance ratio $\sqrt{\frac{\beta_p}{\beta_n}}$. Circuit designers cannot adjust the k' values of their transistors, but they can adjust the transconductance ratio by choosing the W/L s of the pullup and pulldown.

The noise margins are also affected by β_p/β_n. Logic gate reliability must take into consideration the reliability of both on and off devices: the on device in the gate should be very solidly on while the off device should be very solidly off. We can capture this property to some extent using the transfer curve and our analysis of the middle voltage V_M. The middle voltage is a function of the threshold voltages of the transistors. Reducing the threshold voltage lowers the middle voltage, which in turn reduces the gate's noise margin.

3.4 Delay

We now have the pieces in place to analyze the delay through an inverter. We will start with a resistive model of the transistor from which we will build a simple RC delay model [Mea79]. We will then consider the effects of large loads. We will next look at the relationship between logic levels and noise.

3.4.1 Transistor models

To compute the delay through an inverter, we need a model of the transistor's behavior. The transistor equations from Section 2.4.3, while highly simplified relative to the actual device physics, are still too complex and ungainly to use for the design of logic circuits.

We can gain a great deal of insight into the operation of an inverter by considering the transistor to be a resistor. Specifically, we can model it as a combination of a resistor and a switch, as shown in Fig. 3.12. The switch is controlled by the gate voltage—when the gate voltage is below the threshold voltage, the switch is off and the model describes an open circuit. When $V_{gs} > V_t$, the switch is closed and the transistor is modeled as a resistor that connects the source and drain. This model completely ignores the distinction between the linear and saturation regions; it also ignores the fact that current is independent of channel voltage in the saturation regions. Nonetheless, with a properly chosen value for the **effective resistance**, it provides sufficiently accurate results for delay calculations.

As shown in Fig. 3.13, we will compute the transistor's effective resistance using two points on the drain characteristics. We are interested in the curve for which $V_{gs} = V_{DD}$ since that is the maximum voltage at which the transistors will operate in the logic circuit. For the moment, we will assume that $V_{SS} = 0$. We will use the notation $V_B = V_{DD} - V_t$. One point p_{lin} represents the transistor in the linear region; it is chosen at the halfway point between the two ends of the linear region:

$$V_{lin} = \frac{V_{ds}}{2} = \frac{V_B}{2}. \tag{3.11}$$

The final form of this equation comes from two substitutions: the gate is connected to V_{DD} and we assume V_{SS} is zero. We can then compute the current at this voltage either by referring to the curve or by substituting our value for $V_{ds} = V_{lin}$ into the transistor equations:

$$I_{lin} = \beta \left[(V_{gs} - V_t)V_{ds} - \frac{1}{2}V_{ds}^2 \right] = \beta \left[\frac{1}{2}V_B^2 - \frac{1}{8}V_B^2 \right] = \beta \frac{3}{8}V_B^2. \tag{3.12}$$

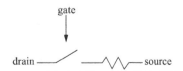

FIGURE 3.12

A simple model of a transistor.

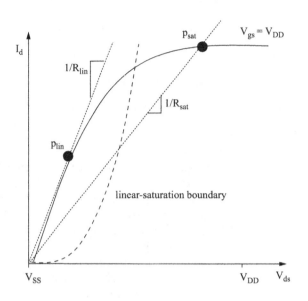

FIGURE 3.13

Resistive approximation of the transistor.

We can also choose p_{sat} at the midpoint of the saturation region to represent that region of operation:

$$V_{sat} = V_{ds} + \frac{V_{DD} - (V_{ds} - V_t)}{2} = V_{DD} - \frac{V_t}{2}. \tag{3.13}$$

$$I_{sat} = \frac{1}{2}\beta(V_{gs} - V_t)^2 = \frac{1}{2}\beta\left(\frac{V_{DD} - V_t}{2}\right)^2 = \beta\frac{1}{8}V_B^2. \tag{3.13a}$$

Since resistance is defined as $R = V/I$, we can calculate the resistance at each point:

$$R_{lin} = \frac{V_{lin}}{I_{lin}} = \frac{4}{3\beta V_B} \tag{3.14}$$

$$R_{sat} = \frac{V_{sat}}{I_{sat}} = \frac{2V_{DD} - V_t}{\beta V_B^2} \tag{3.15}$$

We then use the average of the two as the transistor's effective resistance:

$$R_t = \frac{R_{lin} + R_{sat}}{2} = \frac{10V_B + 3V_t}{6\beta V_B^2}. \tag{3.16}$$

We will use R_t to refer to a generic effective resistance for either an n-type or p-type transistor. When we need to be specific, we will use R_n for n-type and R_p for p-type.

The effective resistance scales with $\frac{L}{W}$ through β: when we double W we cut the transistor's effective resistance in half.

Example 3.1 Transistor Effective Resistance

We can compute the effective resistances of n-type and p-type transistors. The device transconductances and threshold voltages are:

n-type	$k'_n = 200\ \mu A/V^2$	$V_{t,n} = 0.5\ V$
p-type	$k'_p = 50\ \mu A/V^2$	$V_{t,p} = -0.5\ V$

p-type transistors have lower transcondutances because holes have lower effective mobility than do electrons. If we assume that $V_{DD} = 1.2\ V$, $W/L = 1$, then the effective resistances are.

$$R_n = 14.5\ k\Omega$$
$$R_p = 57.8\ k\Omega$$

Example 3.2 Sensitivity of R_t

Variations during manufacturing of the basic device parameters—doping, oxide thickness, etc.—will manifest themselves as variations in higher-level parameters such as R_t. Take as an example a transistor with nominal values $V_t = 0.5\ V$, $k' = 8.1 \times 10^{-4}\ A/V$ that is operated at a power supply voltage of 1 V. A $\pm 20\%$ variation in V_t gives this range of R_t values:

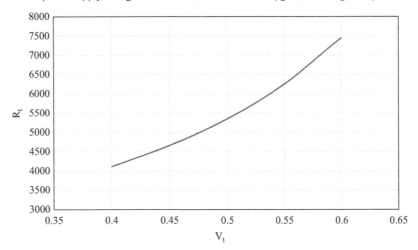

We see an almost $2\times$ variation in effective resistance with only $\pm 20\%$ variation in V_t. A similar $\pm 20\%$ variation in k' gives this result for R_t:

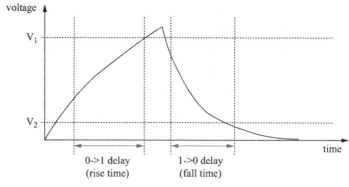

In this case, R_t varies by less than 50%—still significant, but the effective resistance is less sensitive to k' than to V_t.

3.4.2 **RC models for delay**

Waveforms

We can understand delay by measuring it on the output waveforms of an inverter. As shown in Fig. 3.14, we will choose starting and ending voltages from which we can measure the time required for the gate's output to move from the starting value to the ending value. We can measure delay for both the downward ($1 \rightarrow 0$) and upward ($0 \rightarrow 1$) waveforms:

- **rise time** is the time for the output to rise to a logic 1;
- **fall time** is the time for the output to fall to a logic 0.

FIGURE 3.14

Measuring delay from waveforms.

Load and drive

Rise and fall times are not in general equal.

To determine the inverter delay, we need to know what is connected to the output of the inverter. A useful gate has its output connected to another gate. The gate circuit at the output presents a capacitive load to the first gate, as shown in Fig. 3.15. That capacitance comes from the transistor gate capacitance of both the transistors in the second inverter:

$$C_L = C_{g,p} + C_{g,n} \tag{3.17}$$

Fig. 3.16 shows an inverter modeled with switched resistors. We will assume that only one switch is on at any given time. If the input voltage is 0, the p-type is on and the n-type is off; if the input voltage is V_{DD}, then the n-type is on and the p-type is off. This assumption simplifies our circuit to a single resistor and capacitor. This assumption is roughly equivalent to assuming that the gate voltage is a perfect step function.

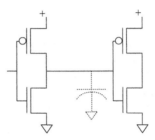

FIGURE 3.15

The capacitive load of an inverter.

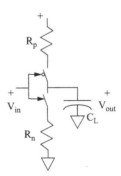

FIGURE 3.16

An inverter modeled with switched resistors.

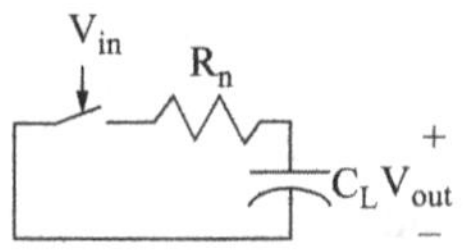

V_{in} R_n

C_L V_{out} $+$ $-$

FIGURE 3.17

RC model of inverter delay.

Fig. 3.17 shows our complete circuit model for inverter delay for the $1 \rightarrow 0$ case. For the $0 \rightarrow 1$ case, the effective resistance is R_p, and a V_{DD} voltage source is inserted between the switch and the capacitor. The more typical model used for RC delay analysis replaces the switch and fixed voltage source with a step voltage source. As Fig. 3.18 shows, our switched resistor model maps more directly onto the inverter.

Pool model

A simple analogy to this circuit model is filling a swimming pool, shown in Fig. 3.19. The quantity of water per unit time flowing through the hose is analogous to current. The height of the water tower that provides the water determines the water pressure, which is analogous to voltage. We can increase the water flow either by using a bigger hose (equivalent to a transistor with a lower effective resistance) or raising the water tower (equivalent to increasing the power supply voltage). The size of the swimming pool that we need to fill is analogous to the load capacitance.

Delay formulas

We will first find an expression for the gate output waveform as a function of time; we will then identify metrics to relate the waveform to delay for discrete output levels. The current through the capacitor and resistor are equal, which we can use to write an equation for the voltages around the circuit:

$$V_{DD} = R_t i(t) + V_c(t) \tag{3.18}$$

The current through the capacitor is

$$i_c = C_L \frac{dV_c}{dt} \tag{3.19}$$

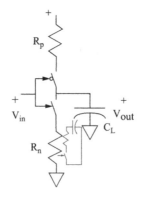

FIGURE 3.18

Relationship between the RC delay circuit and the switched resistor inverter.

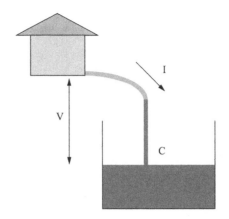

FIGURE 3.19

A water model for delay.

So

$$V_{DD} = R_t C_L \frac{dV_c}{dt} + V_c(t). \tag{3.20}$$

To solve this differential equation, we need to know an initial condition. If we calculate the fall time, then the capacitor starts at $V_c(0) = V_{DD}$. Since the voltage across the capacitor is also the output voltage we are interested in, the output voltage as a function of time is

$$V_{out}(t) = V_{DD}e^{-t/R_t C_L}. \tag{3.21}$$

Fig. 3.20 shows the form of the output voltage, which exponentially decays toward V_{SS}; it never reaches the asymptotic value. The term $R_t C_L$ is known as the **time constant** of the circuit, which is often abbreviated as τ. When $t = R_t C_L$, then the output voltage is $V_{out}(\tau) = V_{DD}e^{-1} = 0.367V_{DD}$.

Delay and rise/fall times   The fact that the output voltage never reaches 0 raises an interesting question: how do we measure delay? Two different definitions of delay are illustrated in Fig. 3.21. One definition, which is often called **delay**, is the timed required for the gate to move from its initial value to 50% of its final value. The other definition, usually

FIGURE 3.20

Output voltage waveform for fall time.

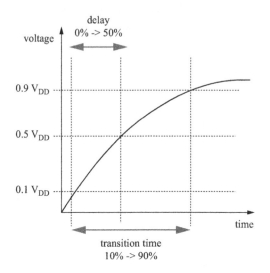

FIGURE 3.21

Definitions of delay.

called **transition time**, refers to the time required for the output to go from 10% of its initial value to 90% of its initial value. (*Transition time* is another generic term; the equivalents for a specific direction are **rise time** and **fall time**). Both definitions have their uses in different situations. Each of these definitions can be used for both rise and fall times.

We can find simple a formula for delay. Starting with our equation for fall time,

$$0.5\ V_{DD} = V_{DD}\left(e^{-t_{0.5}/R_t C_L} - e^{-t_0/R_t C_L}\right) \tag{3.22}$$

Since $t_0 = 0$,

$$t_d = t_{0.5} = -R_t C_L \ln 0.5 = 0.69 R_t C_L \tag{3.23}$$

We can find a similar formula for transition time:

$$t_{rf} = 2.2 R_t C_L. \tag{3.24}$$

Note that we do not directly use our logic levels V_L and V_H. Although we will use them to understand reliability, those levels do not provide good boundaries for evaluating the dynamics of the circuit.

Highlight 3.1

Inverter delay is proportional to $\tau = RC$.

Example 3.3 Inverter Delay

The gate capacitance of one of our minimum-size transistors is $C_g = 0.89$ fF; p-type and n-type transistors of the same size have the same gate capacitance. The total load capacitance is then $C_L = 2C_g = 1.8$ fF. We can then compute the fall times using both delay and transition time metrics, based on the effective resistances of Example 3.1:

Delay (falling)	$t_d = 0.69 \times 14.5$ k$\Omega \times 1.8$ fF $= 18$ ps
Fall time	$t_f = 2.2 \times 14.5$ k$\Omega \times 1.8$ fF $= 57$ ps

We can also compute rise times using the effective resistance of the p-type transistor:

Delay (rising)	$t_d = 0.69 \times 57.8$ k$\Omega \times 1.8$ fF $= 71$ ps
Rise time	$t_r = 2.2 \times 57.8$ k$\Omega \times 1.8$ fF $= 229$ ps

This plot compares the results of a circuit simulation of the inverter and the RC approximation for a falling transition:

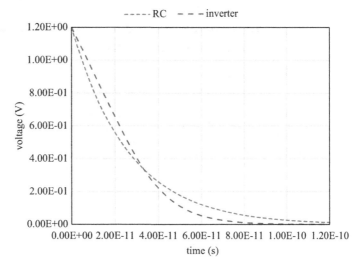

The RC model shows good agreement with the inverter. This simulation used the simple, long-channel model for the transistor. A more sophisticated simulation that takes into account short channel effects would show somewhat less congruence with the RC model, but the RC model is still useful as a hand calculation method.

Transistor sizing

Circuit designers cannot change threshold voltage or transconductance, but they can choose the W/L of each transistor separately. Choosing the sizes of transistors to adjust their delay is known as **transistor sizing**. Effective resistance scales with $1/(W/L)$. We can find the ratio of $\widehat{R}_t$ after sizing to R_t before sizing:

$$\frac{\widehat{R}_t}{R_t} = \frac{\frac{10V_B+3V_t}{6\frac{W}{L}k'V_B^2}}{\frac{10V_B+3V_t}{6\frac{W}{L}k'V_B^2}} = \frac{W/L}{\widehat{W}/\widehat{L}} = \frac{L/\widehat{W}}{L/W} \tag{3.25}$$

so

$$R_t \propto \frac{L}{W}. \tag{3.26}$$

Gate delay is proportional to R_t, so we can reduce gate delay by increasing transistor width.

Highlight 3.2

$R_t \propto \dfrac{L}{W}$

Example 3.4 Transistor Sizing

We are given an n-type pulldown transistor with parameters $V_{tn} = 0.5$ V, $k'_n = 80$ µA/V2. It is part of an inverter with $V_{DD} = 1.2$ V, $C_L = 5$ fF. If we let $W/L = 1$, $R_{n1} = \frac{10V_B+3V_t}{6\beta V_B^2} = 36.1$ kΩ. We can rewrite the fall time formula as a function of transistor size:

$$t_f = 2.2\frac{R_{n1}}{\left(\frac{W}{L}\right)}C_L.$$

We can plot fall time for a range of transistor sizes:

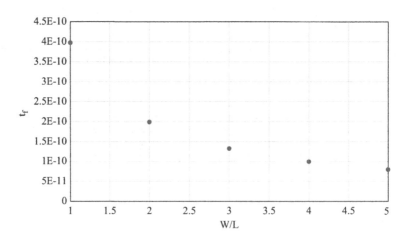

The lower effective mobility of holes has some strong consequences: the lower current results in a higher effective resistance, which makes the rise time substantially longer than fall time if the p-type transistor is the same size as the n-type transistor. The solution is to make the p-type transistor wider. We want the condition

$$t_r = t_f, \tag{3.27}$$

$$2.2R_pC_L = 2.2R_nC_L. \tag{3.28}$$

This requires that the effective resistances of the n-type and p-type be equal. Since $R_t \propto 1/\beta$, the condition $R_p/R_n = 1$ implies

$$\frac{(W/L)_p}{(W/L)_n} = \frac{k'_n}{k'_p}. \tag{3.29}$$

3.4.3 Drive and loads

Another simple way to think about a gate's delay is to consider it as a current source. When either of the transistors is in saturation, its output current is independent of the drain-to-source voltage, making it to operate as a current source. Since we use the current to charge the capacitance at the inverter's output, more current means that we will charge the output capacitance faster. Increasing the size of the n-type decreases t_f, and increasing the p-type's size causes t_r to decrease.

Unfortunately, as we give one gate more drive, it also presents a larger capacitance at its input, slowing down the previous gate. As shown in Fig. 3.22, increasing W/L increases the gate's C_g, which in turn increases the load presented to the previous gate. Making the previous gate larger only pushes the problem back one level—it does not solve the problem.

An extreme form of this problem occurs when we try to communicate off-chip. The real world does not scale; as Moore's Law progresses, the drive current of minimum-size transistors decreases while the capacitance at the chip's edges (the

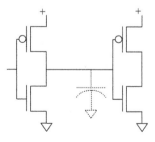

FIGURE 3.22

Drive and capacitive load.

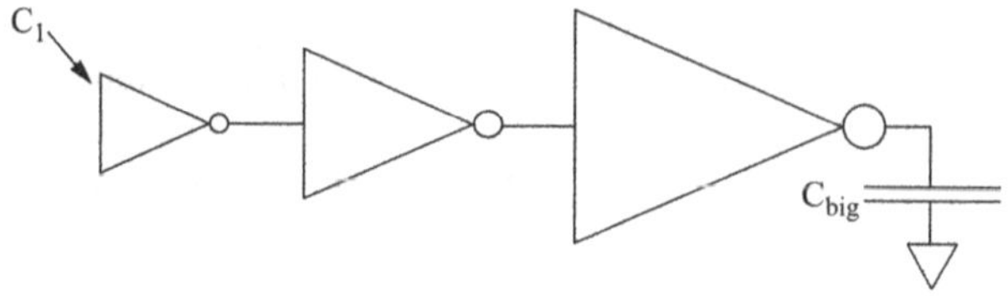

FIGURE 3.23

Driving large loads through a chain of drivers.

printed circuit board, for example) stays the same. While we could use large transistors at the chip's outputs to drive the external loads, this would simply create large loads for earlier stages of logic. The best way to handle this problem, as illustrated in Fig. 3.23, is with a cascaded series of inverters. Each presents a somewhat larger load to the previous stage and provides more drive for its load.

Driver chains   We need to determine both the number of stages required and how to size the transistors in each stage of the cascade [Jae75]. We will assume that the ratio of drive capability is the same at every stage i:

$$\frac{W_{i+1}}{L_{i+1}} = \alpha\frac{W_i}{L_i} \tag{3.30}$$

We want to minimize the total delay through the driver cascade. Our assumption of equal α at every stage allows us to spread the delay evenly among the stages. If t_1 is the delay of the first stage, C_1 is the capacitance of the first inverter in the chain, and C_{big} is the capacitance at the output of the last stage, then the delay through the cascade of n inverters is

$$t_c = n\left(\frac{C_{big}}{C_1}\right)^{1/n} t_1 \tag{3.31}$$

We can find the minimum number of buffers by minimizing the derivative of delay:

$$\frac{dt_c}{dn} = n\left(\frac{C_{big}}{C_1}\right)^{1/n} t_1 = 0 \tag{3.32}$$

This gives

$$n = \ln\frac{C_{big}}{C_1} \tag{3.33}$$

When we use this value of n in the delay formula, we find that

$$\alpha = e. \tag{3.34}$$

Thus, the transistor sizes in the inverter cascade are **exponentially tapered**. This is an example of **impedance matching**, a basic phenomenon in circuit design.

3.5 Power and energy

We would also like to know the energy and power consumed by a gate. Ideal CMOS logic revolutionized electronics by enabling a new class of portable, battery-operated device. The TRS-80 Model 100, shown in Fig. 3.24, was the first computer to use a CMOS microprocessor. It was for many years the best-selling computer in the world.

The CMOS microprocessor in the TRS-80 Model 100 competed against microprocessors built with nMOS logic. Fig. 3.25 shows an nMOS inverter. The pullup transistor is an enhancement mode transistor that is doped to be normally on; the pullup's gate is biased so that the depletion transistor acts as a resistor. When the input goes low, the pulldown transistor is off, and the resistor charges the output capacitance with the output current exponentially decreasing. When the input goes high, the pulldown transistor is on and forms a resistive divider with the pullup. In this case, the output voltage never goes to zero, and current always flows through the pullup and pulldown. As a result, the nMOS inverter consumes a significant amount of static power. CMOS gates do not use energy when their outputs are not transitioning because the pullup and pulldown are not on at the same time.

Switching energy   The form of the formula for energy consumption of a gate has a remarkably simple form. A transition occurs by charging or discharging a capacitor. The energy for charge a capacitor is $= \frac{1}{2}CV^2$, a fact we can derive by integrating the potentials of a differential charge added to the capacitor:

$$E_c = \int_0^Q V(q)dq = \int_0^Q \frac{q}{C}dq = \frac{1}{2}\frac{Q^2}{C} = \frac{1}{2}CV^2. \tag{3.35}$$

FIGURE 3.24

The TRS-80 Model 100 portable computer.

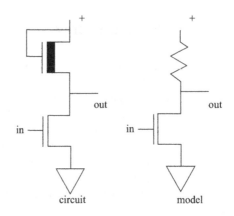

FIGURE 3.25

An nMOS inverter.

The same amount of energy is required to discharge the capacitor. We can simplify the formula for gate energy consumption by defining the unit energy consumption of a gate as the energy required for one rise and one fall. The gate's **switching energy** consumption E_s is then

$$E_s = E_{c,r} + E_{c,f} = C_L V_{DD}^2. \tag{3.36}$$

Remarkably, this formula depends on the load capacitance at the output but not the resistance of the transistor driving the output. However, it does depend on the output capacitance, which is determined by the size of the transistors in the next gate. We will see later that increasing the size of transistors in a gate increases the load on the previous gate, potentially slowing it down.

Short circuit current   Our assumption that only one transistor is on at a time is not strictly true. During a transition, both transistors are on for a brief period resulting in **short circuit current**. As shown in Fig. 3.26, not all of the current going through the transistors goes to charge the output capacitance [Aga07]:

$$I_{dp} = I_{dn} + I_L \tag{3.37}$$

As illustrated on Fig. 3.27, the gate conducts short circuit current in the region for which both pullup and pulldown transistors are above their threshold voltages. The short circuit current can be made larger by any of several factors:

- A smaller load capacitance means that less current is required for the capacitor and more of the current produced by the transistors must be shunted to the short circuit path.
- A higher transistor transconductance produces more current, some of which may not be needed to charge or discharge the capacitance.
- Slower input transitions mean that the transistors spend more time in the short circuit region of the transfer curve.

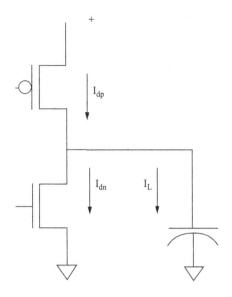

FIGURE 3.26

Short circuit current in the inverter.

Because short circuit current requires that both transistors are on, it goes to zero if the threshold voltages are low enough such that

$$V_{DD} < V_{tn} + |V_{tp}| \tag{3.38}$$

The capacitors that form the load capacitances also leak charge slowly over time, resulting in some power consumption.

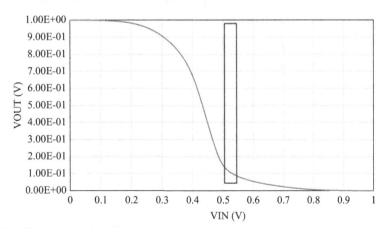

FIGURE 3.27

The short circuit region on the transfer curve.

Dynamic energy

The total energy used by a gate to perform logical operations is known as **dynamic energy** E_d. Switching energy and short circuit energy are the two major components of E_d.

Power

Power is energy per unit time. We can find a simple formula for power consumption if we assume that we know the amount of time required for a transition. We will call this value f; as we will see, this frequency is determined by the operating speed of the logic. The switching power consumption of a gate is then

$$P_s = f C_L V_{DD}^2. \tag{3.39}$$

This formula does not directly reference the effective resistance, but f in fact depends on the transistors' effective resistances. This relationship shows the trade-off between delay and energy—to make gates run faster, we must increase their power consumption.

Delay versus power

How does delay vary with power? We can develop a basic relation by analyzing the effective resistance of the transistor at the end of its range: when both the gate and source/drain voltages are at the power supply.

$$R_{DD} = \frac{V_{sat}}{I_{sat}} = \frac{V_{DD}}{1/2\beta(V_{DD} - V_T)^2} \propto \frac{1}{V_{DD}} \tag{3.40}$$

We can see that effective resistance, and therefore RC delay, scales linearly with power supply voltage.

A convenient metric for the efficiency of logic is the **speed–power product**:

$$SP = \frac{1}{f}P = C_L V_{DD}^2. \tag{3.41}$$

Highlight 3.3

$$P_d = f C_L V_{DD}^2, SP = \frac{1}{f}P = C_L V_{DD}^2$$

Total power

Many logic families consume energy even when their values are not changing, known as **static power**. The gate's total power consumption is the combination of dynamic and static power:

$$P_g = P_d + P_s \tag{3.42}$$

While early CMOS logic consumed very little static power, modern CMOS devices make use of very small, imperfect devices that consume power even when the gate's output is not changing such as the leakage mechanisms we discussed in Section 2.4.4.

We cannot write a simple formula for leakage power as we did for dynamic energy and power—leakage depends on both device and circuit characteristics.

Leakage control

We can eliminate a gate's leakage current by removing it from the power supply. We can add circuitry to dynamically connect and disconnect the logic gate to the power supply during operation. As shown in Fig. 3.28, a transistor is used as a power management switch. The transistor used is a special, high-threshold transistor that must be added to the manufacturing process. External power management logic controls the *sleep* signal; power management logic can often determine that a given block of logic will not be used for a given interval, allowing it to safely turn off the logic and reduce leakage.

DVFS

This simple metric suggests that we can trade off power consumption and performance using the power supply voltage. When we combine the speed–power product with EQ 3.40, we come to a surprising conclusion: delay increases linearly with power supply voltage but power consumption decreases quadratically with V_{DD}. This means that if we can tolerate decreased performance, we can achieve large gains in power consumption. Modern computer systems constantly perform **dynamic voltage and frequency scaling (DVFS)** to manage power consumption. Not all applications require maximum performance. The performance requirements on a system can even change over time—a laptop's workload can vary drastically depending on the applications run by the user. Modern processors and operating systems evaluate the performance requirements of the system, then adjust the power supply voltage and the clock frequency to the lowest levels required to meet the current performance demands.

Race-to-dark

Even with careful circuit design, leakage currents remain a substantial problem. Leakage power consumption is larger than dynamic power consumption in many

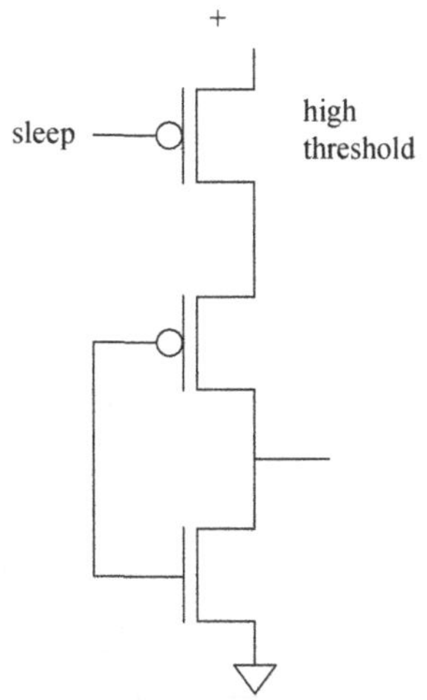

FIGURE 3.28

Power management for logic gates.

nanometer-scale manufacturing technologies. When leakage power consumption is a major problem, computer systems use a different power management policy known as **race-to-dark**. While DVFS slows down the processor to just meet the minimum required performance, race-to-dark runs as fast as possible and shuts down the logic as soon as the task is finished. Many smartphones use race-to-dark power management.

3.6 Scaling theory

Moore's Law is purely descriptive—it observes that transistor sizes are decreasing at an exponential rate. However, the effect of smaller transistors on logic circuits is not immediately clear. While scaling transistor sizes allows more transistors per chip, it is not at all clear whether logic circuits become slower or faster with scaling. The effects of scaling on energy consumption are equally important. In 1974, Peter Dennard and colleagues at IBM wrote a paper on advanced MOS digital circuits [Den74]. Part of their paper proposed a model for how circuit properties change as technologies scale. They showed a very surprising result: scaling actually makes gates run faster (although wires become slower).

Scaling model   We are interested in comparing successive generations of fabrication technology. A new generation of technology is **scaled** by $1/x$ compared to the previous generation. Fig. 3.29 shows Dennard's model for transistor scaling. All of the geometric measurements are scaled by $1/x$: the transistor length and width as well as the gate oxide thickness. We also scale the power supply voltage by $1/x$. To keep the device parameters at reasonable levels, we must increase the doping. This gives us several scaling relationships:

$$\widehat{W} = W/x \tag{3.43}$$

$$\widehat{L} = L/x \tag{3.44}$$

$$\widehat{N_d} = N_d x \tag{3.45}$$

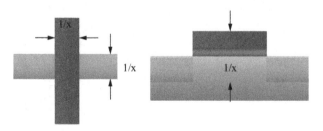

FIGURE 3.29

Dennard et al.'s model for transistor scaling.

$$\widehat{V_{DD}} = V_{DD}/x \tag{3.46}$$

Given these relationships, we can determine how gate capacitance scales:

$$C_g = \frac{\varepsilon_{ox} WL}{t_{ox}} \tag{3.47}$$

$$\frac{\widehat{C_g}}{C_g} = 1/x \tag{3.48}$$

We can also compute how saturation drain current scales:

$$\frac{\widehat{I_d}}{I_d} = \frac{\widehat{k'}}{k'} \frac{\widehat{W}/\widehat{L}}{W/L} \frac{\left(\widehat{V_{gs}} - \widehat{V_t}\right)^2}{(V_{gs} - V_t)^2} = \frac{1}{x}. \tag{3.49}$$

Gate delay scaling

Dennard et al. modeled gate delay as

$$t = \frac{CV}{I}. \tag{3.50}$$

This is a reasonable approximation of the RC gate delay. We can use this definition to determine how performance scales:

$$\frac{\widehat{t}}{t} = \frac{\widehat{C}\widehat{V}/\widehat{I}}{CV/I} = \frac{1}{x}. \tag{3.51}$$

This result means that gates actually get faster with scaling. Fig. 3.30 helps explain why the current delivered by each transistor scales as $1/x$. The load that each transistor drives, however, depends on the area of the gate capacitance, which decreases more quickly as $1/x^2$. Since the load goes down faster than the drive decreases, gates run faster.

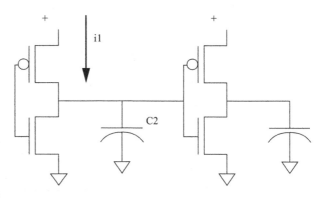

FIGURE 3.30

Loads and drive currents.

Power scaling

Power consumption also improves with scaling. Since $P = VI$, power scales as $1/x^2$. The speed–power product scales as $1/x^3$. The **power density** D_P, power per unit area, stays constant with scaling. If power density increased, then chips would burn more and more power per unit area with scaling, which would cause a variety of problems. Power density does not improve with scaling but it also does not get worse.

Wire scaling

The behavior of wires is less desirable. Fig. 3.31 shows Dennard et al.'s model for wire scaling: the length, width, and height of the wire all scale as $1/x$. Dennard used wire resistance as a metric for wire delay. Remember that the resistance of a wire is given by

$$R = \rho \frac{L}{A}. \tag{3.52}$$

In this formula, ρ is the resistivity of the material while L and A are the length and area of the wire, respectively. Resistance scales as

$$\frac{\widehat{R}}{R} = \frac{\widehat{L}/\widehat{A}}{L/A} = x. \tag{3.53}$$

This means that wire resistance grows with scaling. Wire current density I/A grows as x, which means that wires are subjected to larger currents with scaling, increasing the potential for damage as wires with smaller and smaller dimensions take larger and larger currents. We will see later that Dennard's results for wire delay were actually optimistic—a more detailed model shows that delay actually is a function of the square of the length of the wire.

Even worse, scaling theory predicted that wire delay would increase relative to gate delay since the gates' RC time constant scales as $1/x$. This prediction has, in fact, been fulfilled—wire delay is greater than gate delay in many logic designs.

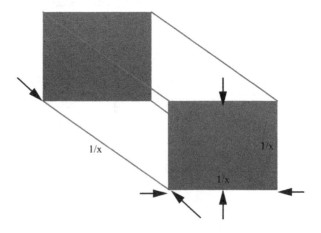

FIGURE 3.31

Dennard's model for wire scaling.

We will see that a variety of factors have caused Dennard's scaling model to break down in recent years. The failure of ideal scaling has created substantial new problems in power consumption and heat generation.

Example 3.5 Scaling in Practice

How has scaling worked in practice? Here are actual values for performance for microprocessors over several generations of technology from the ITRS Roadmaps:

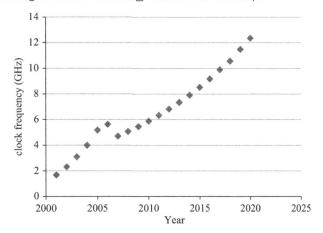

Frequency, which is related to gate delay, has risen exponentially, as classical scaling predicted. However, power consumption has increased markedly:

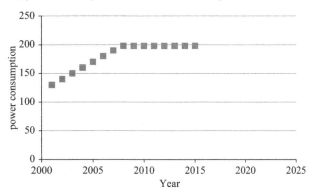

These data are for high-end microprocessors; chips for low-power applications have seen similar trends. Power consumption is limited in large part by our ability to remove heat from the package.

Since power has increased faster than chip area, power density has increased, contrary to classical scaling. We will see later the physical effects that have contributed to increasing power consumption.

We can also see that power supply voltage has not scaled down in recent years as predicted:

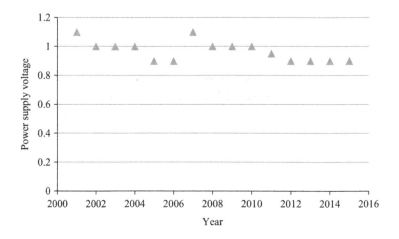

3.7 **Reliability**

We want our logic gates to operate reliably. We often assume that our computers oper-ate error-free but that is not the case. We saw in Section 2.3.2 that randomness is a fundamental aspect of the behavior of matter.

Barriers   One basic definition of reliable operation of a digital circuit is that its output does not change value unexpectedly. Thermal energy always excites electrons to some extent; we want the logic to be designed such that it is very unlikely that enough elec-trons will gain enough thermal energy (or energy from other outside sources) so as to move from where they should be to where we do not want them to be. As shown in Fig. 3.32, the energy bands of a MOSFET in cutoff act as **energy barriers**. An elec-tron moving between the source and drain while the transistor is off would be a source of noise. For an electron to do that, it has to have enough energy to overcome the en-ergy difference between the bands in the source/drain versus the channel.

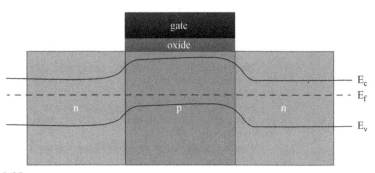

FIGURE 3.32

Energy bands in the transistor as barriers.

Errors

Although integrated circuits have many sources of noise, we will concentrate here on thermal noise. We can represent the probability of noise reaching given level as [Lan61; Key70]

$$P_{err} = e^{-E_b/kT} = e^{-qV/kT} \tag{3.54}$$

This formula is based on the exponential distribution of electron energies that we saw in Chapter 2.

Highlight 3.4

Error rates due to thermodynamic noise can be modeled as $P_{err} = e^{-E/kT}_b$.

Fundamental limits

We can use a worst-case scenario to examine the question of the energy used to represent a bit. We will assume that we use one electron to represent a bit. To provide a very optimistic estimate of the minimum energy required to store a bit, let us use a probability of error for a bit as $P_{err} = 0.5$, or a 50-50 chance of a bit changing value. While this probability of error would be unacceptably high in a real system, it does provide a minimal standard for us to evaluate the energy required to represent bits. In this case, the minimum bit energy is [Key70]:

$$E_b = kT \ln 2 = 0.7kT \tag{3.55}$$

Example 3.6 Error Rates and Power Consumption

We can compare the error rates of our worst-case model to a more typical case. At room temperature the worst-case model has an energy per bit of

$$E_b = 2.8 \times 10^{-21} \text{ J}$$

For comparison, how much energy does it take to change the value of 1 bit in a typical inverter? This is equivalent to finding the amount of energy stored by the output capacitance of the gate. Assume that $C_L = 1.8$ fF. Since the energy required to charge or discharge a capacitor is $1/2CV^2$, then the energy required to change 1 bit at a power supply voltage of 1 V is

$$E_b = \frac{1}{2}(1.8 \text{ fF})(1 \text{ V})^2 = 0.9 \text{ fJ} \tag{3.56}$$

This energy is about 10^5 higher than our idealized case. At these energy levels, $P_{err} = e^{-0.9 \times 10^{-15}/kT} \approx 0$.

We can also estimate chip power consumption from these E_b values. Assume that our computer has $n_d = 1 \times 10^9$ devices and runs at a rate of $=1 \times 10^{-9}$ s. The total chip power consumption is $E_b n_d f = 0.0028$ W of power. This is several orders of magnitude smaller than the actual power consumption of modern high-end chips, which ranges between tens and hundreds of watts. However, this chip would also be completely useless: half of its billion devices would fail on every cycle, which would occur one billion times per second.

In contrast, our capacitor model predicts that the chip would consume 900 W. This figure is somewhat high—a server CPU may burn a few hundred watts. However, not all of the transistors in a realistic CPU change state on every clock cycle. When logic activity is factored in, the estimate moves closer to actual power consumption values.

Some theoretical studies have proposed computational models that use even less energy than the $kT\ln2$ limit. Bennett [Ben73], for example, showed how to formulate a reversible Turing machine. By using operations that can be reversed, the machine can avoid dissipating energy. Quantum computers are one effort to find physical machines that can perform these low-energy reversible computations. We will discuss this approach in Chapter 7.

3.8 Synthesis

- Performance, energy/power, and reliability are three critical characteristics of logic gates.
- Static analysis tells us how to choose logic levels that determine the relationship between voltages and Boolean values.
- Logic gates are amplifiers.
- Logic gate performance can be characterized by rise/fall time, which can be estimated using RC models.
- Dynamic energy consumption of a logic gate is the current consumed by the gate while changing its output. Dynamic energy consumption does not directly depend on the sizes of the driving transistors, only the size of the capacitive load and power supply levels.
- Static energy consumption is driven by leakage, with subthreshold conduction being a major component of leakage current.
- Logic gate reliability is influenced by a number of factors. Making gates more reliable generally requires spending more energy.

Questions

Q3-1   Assume $k' = 150\ \mu A/V^2$, $V_t = 0.5$ V, $W/L = 1.5$, $V_{gs} = V_{DD} = 1.2$ V. Find the drain current at two points:

   a. The midpoint of the saturation region.
   b. The midpoint of the linear region.

Q3-2   Compute the effective resistances of these transistors (assume $W/L = 1$):
   a. n-type $V_{tn} = 0.5$ V, $k'_n = 80\ \mu A/V^2$, $V_{DD} = 1$ V.
   b. n-type $V_{tn} = 0.4$ V, $k'_n = 200\ \mu A/V^2$, $V_{DD} = 1.2$ V.

c. p-type $|V_{tp}| = 0.4$ V, $\left|k_p'\right| = 35$ µA/V^2, $V_{DD} = 1$ V.

d. p-type $|V_{tp}| = 0.55$ V, $\left|k_p'\right| = 80$ µA/V^2, $V_{DD} = 1.2$ V.

Q3-3 You are given a gate with $R_n = 6.5$ kΩ, $C_L = 0.9$ fF, $V_L = 0.55$ V, $V_H = 0.65$ V, $V_{DD} = 1.2$ V.

a. Plot the waveform of the high → low transition for the RC circuit in the time range [0,18 ps].

b. Over what range of times is the gate's output X?

Q3-4 You are given a gate with $R_p = 35$ kΩ, $C_L = 4$ fF, $V_L = 0.25$ V, $V_H = 0.7$ V, $V_{DD} = 1$ V.

a. Plot the waveform of the low → high transition for the RC circuit in the time range [0,350 ps].

b. Over what range of times is the gate's output X?

Q3-5 Derive a formula for delay using the 20−80% definition of delay.

Q3-6 Assuming that the resistance of a minimum-width transistor is 6.5 kΩ and the capacitance of a minimum-width transistor is 0.9 fF, plot the RC 10−90% delay of an inverter as its W/L varies from 1 to 5 under these conditions:

a. Load capacitance is constant at 1.8 fF.

b. Load capacitance is equal to 2× the gate capacitance of the driving transistor.

Q3-7 You are given a transistor whose effective resistance 6.5 kΩ. Plot the RC 10−90% delay of an inverter as its load capacitance varies from 1.8 to 9 fF.

Q3-8 You are given a transistor with an effective resistance of $R_n = 10$ kΩ. The gate's power supply is 1 V. Plot the energy required for a 0 → 1 transition as its load capacitance varies from 2 to 10 fF.

Q3-9 Compute the RC delay for an inverter with $R_n = 20$ kΩ, $R_p = 85$ kΩ, $V_{DD} = 1.0$ V, and total load capacitance of 2 fF:

a. Rise time.

b. Fall time.

c. 0−50% rising delay.

d. 0.50% falling delay.

Q3-10 Plot fall time for these parameters as a function of transistor size for $1 \leq \frac{W}{L} \leq 5$.

a. $V_{tn} = 0.5$ V, $k_n' = 80$ µA/V^2, $V_{DD} = 1$ V, $C_L = 3$ fF.

b. $V_{tn} = 0.6$ V, $k_n' = 200$ µA/V^2, $V_{DD} = 1.2$ V, $C_L = 3$ fF.

Q3-11 Plot rise time for these parameters as a function of transistor size for $1 \leq \frac{W}{L} \leq 5$.

a. p-type $|V_{tp}| = 0.5$ V, $\left|k_p'\right| = 35$ µA/V^2, $V_{DD} = 1$ V, $C_L = 4$ fF .

b. p-type $|V_{tp}| = 0.6$ V, $\left|k_p'\right| = 60$ µA/V^2, $V_{DD} = 1.2$ V, $C_L = 5$ fF.

Q3-12 You are given parameters for an inverter's pullup and pulldown transistors, load capacitance, and power supply voltage. If the n-type $W/L = 1$, what p-type W/L is required to make the rise time at least as fast as the fall time? Round to the nearest integer W/L.

    a. $V_{DD} = 1$ V, $C_L = 4$ fF, n-type $V_{tn} = 0.5$ V, $k'_n = 150\ \mu\text{A}/\text{V}^2$, p-type $|V_{tp}| = 0.5$ V, $\left|k'_p\right| = 35\ \mu\text{A}/\text{V}^2$,

    b. $V_{DD} = 1.2$ V, $C_L = 5$ fF, n-type $V_{tn} = 0.55$ V, $k'_n = 120\ \mu\text{A}/\text{V}^2$, p-type $|V_{tp}| = 0.55$ V, $\left|k'_p\right| = 40\ \mu\text{A}/\text{V}^2$,

Q3-13 You are given transistors with $V_{tn} = 0.45$ V, $V_{tp} = -0.5$ V, $k'_n = 140\ \mu\text{A}/\text{V}^2$, $k'_p = 30\ \mu\text{A}/\text{V}^2$. Compute V_M for an inverter using these transistors with a 1 V power supply and $W/L = 1$.

Q3-14 You are given a driver with a capacitance of 2 pF. What is the optimal load your circuit can drive with an optimal buffer chain of six stages?

Q3-15 Plot the speed–power product for the gate with $R_n = 6.5$ kΩ, $R_p = 15$ kΩ, $C_L = 2$ fF over a range of power supply voltages from 0.7 to 1.2 V.

Q3-16 Plot Dennard scaling for six generations of technology, each of which scales as ½. In the first generation, assume delay is 20 ns and power is 10 μW.
    a. Delay, generation 1 delay is 20 ns.
    b. Power, generation 1 power is 10 μW.

Q3-17 Why did Dennard model gate delay as CV/I? How does that model compare to the exponential RC model?

Q3-18 A given technology has an ideal scaling parameter of $t = 50$ ps. If successive technology generations scale at the rate of $1/x = 1/1.5$, what will be the ideally scaled delay after three generations of scaling?

Q3-19 Minimum-size transistors were 12 μm long in 1974 gate delay was 1 μs, and a chip consumed 1.3 W. After 25 generations with scaling of $x = 2$ at each generation, what should be a chip's gate delay and power consumption?

Q3-20 A bit is stored on a 30 fF at 1 V. What is the mass of this bit?

Sequential Machines

4

4.1 Introduction

In this chapter, we will build sequential machines that provide the next layer of abstraction in the development of computers, namely discrete time. We will start by examining the performance, energy consumption, and reliability of the blocks of combinational logic that are at the heart of a sequential machine. Section 4.3 will consider the behavior of wires, whose nonideal characteristics pose significant challenges. Section 4.4 develops the design of sequential machines that operate on sequences of inputs.

4.2 Combinational logic

We need more than a single gate to build interesting machines. Networks of gates allow us to build complex functions. **Combinational logic** performs Boolean functions but it has no memory. We will start by considering basic models for combinational behavior and structure in Sections 4.2.1 and 4.2.2. Sections 4.2.3 and 4.2.4 study gain and its relationship to both reliability and delay. Section 4.2.5 considers the relationship between delay and power in logic networks. We will introduce the concept of signal integrity in Section 4.2.6. Section 4.2.7 looks at the effects of power supply noise on reliability. Section 4.2.8 considers the effects of coupling between a gate's input and output.

4.2.1 The event model

The use of discrete values to represent voltage waveforms gives us a powerful abstraction for time—the **event model**. Since a range of voltages represents a logic 0 or 1, we do not need to worry about the exact voltages on the gates. We can concentrate on when those values move between our discrete values of 0, 1, and X. We call a change in the discrete value of a signal an **event**. We often refer to an event as a **transition.** An event is modeled by a value/time pair:

$$E = \langle v, t \rangle \tag{4.1}$$

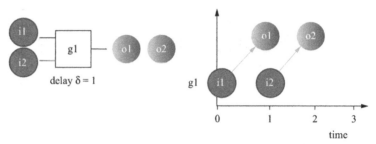

FIGURE 4.1

Events on a gate.

As shown in Fig. 4.1, we can think of the operation of a gate in terms of events. If a gate's inputs do not change their discrete values, in normal operation its output will not change. When an input event occurs, it generates an output event. We take delay into account when we determine the time of the output event. If the delay of the gate is δ, then an input event at time t generates an output event at time $t + \delta$:

$$E_i = \langle v_i, t \rangle \rightarrow E_o = \langle v_o + t + \delta \rangle \tag{4.2}$$

The time between the input and output event can be represented by our delay model, such as inertial or transport delay.

Gate delay models   We often want to determine delay without reference to detailed waveforms. Two logic-level delay models, **inertial** and **transport delay**, are commonly used. Both assume that delay is a value independent of the exact form of the waveform. But they differ in a key respect. Inertial delay models assume that **glitches** do not cause any change in the output. If, for example, the gate's input briefly changes from 1 to 0 and then back to 1, if the duration of the glitch is below a given value, the gate's output is assumed not to change at all. Transport delay, in contrast, propagates even short glitches.

4.2.2 The network model

To build interesting systems, we need to connect together gates into **combinational logic networks**, as shown in Fig. 4.2. We now think of the input and output values at the gates as 0s and 1s. We refer to the gate inputs that are not fed by other gates as **primary inputs** and the gate outputs that do not connect to other gates as **primary outputs**. Our discrete representation for the values of logic gates now allows us to deal with the values in combinational logic more abstractly.

We model the structure of a combinational network as a graph. We can use several different types of graphs, depending on the application. For delay analysis, we often use a directed graph:

* The nodes N in the graph are the combination of the logic gates $g \in G$, the primary inputs $i \in I$, and primary outputs $o \in O$.

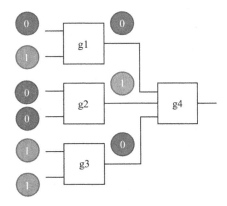

FIGURE 4.2

Discrete values in combinational logic.

- A directed edge $n_1 \rightarrow n_2$ indicates that signals flow from node n_1 to node n_2. We refer to these nodes as the **source** and **sink** of the edge.

For functional analysis and some types of delay analysis, we need more detail:

- The nodes N in the graph include the inputs and output pins on the logic gates $g_i \in G_i$, $g_o \in G_o$, the primary inputs $i \in I$, and primary outputs $o \in O$.
- A directed edge $n_1 \rightarrow n_2$ indicates that signals flow between node n_1 to node n_2.

We often refer to the structure of a logic network as a **netlist**. We will assume that our combinational logic networks are **acyclic** with no path from a gate's output to any of its inputs.

The event model makes it much easier to understand the behavior of a complex network of gates. As shown in Fig. 4.3, events **propagate** through the network from inputs to outputs—events at the inputs of gates cause them to generate new events that stimulate the inputs of the next gates. We can determine all the changes in logical value in the network simply by tracing events through the logic. Events cause a cascade of events in the logic: an event at one of the primary inputs causes

Delay

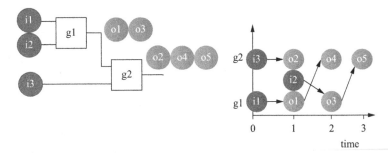

FIGURE 4.3

Events in a combinational network.

Parallelism

an event at the output of its associated gate; that event causes a change at the output of the next gate, and so on, until the event reaches a primary output.

Combinational logic performs all of its operations in parallel—all gates run simultaneously. A C language program, in contrast, runs entirely sequentially. Even most parallel programming languages rely on some amount of sequential behavior. The highly parallel nature of digital logic is the source of its computational power. But parallelism also makes logic harder to design and debug.

Delay

Our primary delay metric is the **worst-case delay** through a block of combinational logic—the longest delay from any input to any output. This worst-case determines the overall performance of the machine. As we will see in Chapter 5, worst-case delay determines clock period.

Consider the logic network of Fig. 4.4. We build a delay model graph: each logic gate node is labeled with the delay from an input to its output; the nodes for primary inputs and outputs have 0 labels; the edges indicate signal flow.

The worst-case delay is determined by the graph's **critical paths**. The critical paths are the longest paths from any input to any output; a network may have more than one such path. A simple way to find the critical paths is to compute the forward and backward paths. First, label each primary input as 0. Then walk along the edges from source to sink starting from the primary inputs. Assign each node n a distance d:

$$d(n) = \max_{c \,\epsilon\, fanin(n)}(d(c)) + w(n) \tag{4.3}$$

The *fanin(n)* function returns all the nodes that are the sources of edges that terminate on n. Next, set the primary output values to the largest such value found in the forward search; walk backward through the graph and assign distances from the primary outputs. Any primary input with a 0 backward weight is on the critical path; any gate with the same distance value on both the forward and backward paths is also on the critical path. In the example, the critical path includes both inputs of B, the bottom input of C, and the top input of E.

The critical path helps us to identify which gates need to be sped up to decrease the delay of a logic network. If we want to speed up the logic, we must speed up gates that form a **cutset** of the critical path—a set of edges that, when removed, cuts all paths from the primary inputs to the primary outputs. If the critical path

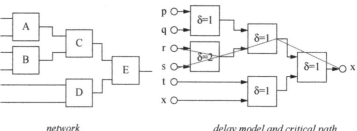

network   delay model and critical path

FIGURE 4.4

Critical delay paths.

has multiple branches, speeding up only one branch still leaves the other branch to cause a critical delay.

How do we assign delay values for gates? The simple approach would be to use the worst-case delay for each gate. We quickly run into several limitations. As we saw in Section 3.4.2, gate delay depends on the output load. A pessimistic assumption would use the worst-case output capacitance. As we saw in Chapter 2, p-type transistors produce less current than equivalent-sized n-type transistors. The rise and fall times of a gate are often not symmetric. Keeping track of values means that we have to find the set of inputs that produces the worst-case behavior; this is an intractable problem. In a few cases, we can show that certain combinations of rising and falling signals cannot occur. The canonical case is a chain of inverters [McW80]: when the output of the first inverter falls, the second rises. The cascaded inverter outputs cannot both rise and fall at the same time. Designers who need a highly accurate analysis of timing for critical blocks of combinational logic generally use a circuit simulator to capture the range of physical phenomena that affect delay and their interactions.

Simulation

We can also use the network model to simulate the function of the network. In this case, the nodes are modeled as Boolean functions. By applying values to the primary inputs, we can walk through the network and find the values at the primary outputs. Many simulators combine functional models with timing models to determine when signals change as values propagate through the network.

4.2.3 Gain and reliability

Logic gates are amplifiers. This has been true from the early days of vacuum tube computers. Amplification consumes energy but helps us to ensure the integrity of our computations.

Gain is the quantitative representation of amplification. Gain is in general a relationship between input and output levels. For example, voltage gain is

$$A = \frac{V_{out}}{V_{in}} \tag{4.4}$$

(A is the traditional symbol for gain, as in *amplification*.) We can read gain from the transfer curve—as shown in Fig. 4.5, gain is the slope of the transfer curve. The negative value of slope in the inverter's transfer curve indicates that the inverter is an inverting amplifier. The gain of an inverter varies across its operating range; we are typically interested in its gain near the center of the transfer curve. We can control the voltage gain of the inverter by adjusting the W/L of the two transistors. If we consider the drain current equation, increasing the transistor width means that we can provide the same current at a lower value of V_{ds}.

A simple model of the inverter is a **saturating inverting amplifier**. Fig. 4.6 compares the input and output voltages of the amplifier. Negative gain corresponds to the inverting nature of the amplifier: a low input voltage produces a high output

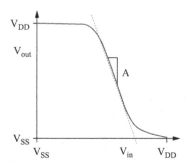

FIGURE 4.5

Inverter gain from the transfer curve.

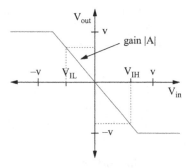

FIGURE 4.6

Gain and logic levels.

voltage. For simplicity, we will center the output voltage axis around the midpoint between the power supply rail, which we will refer to as $\pm v$. This gives us:

$$V_{out} = -AV_{in}, -v \leq V_{in}/_A \leq v$$

$$= v, V_{in}/_A < -v$$

$$= -v, v < V_{in}/_A \tag{4.5}$$

The amplifier's output saturates when it reaches the power supply levels. We can see how the amplifier's gain translates V_{IL} and V_{OL} to output levels.

Fig. 4.7 shows how different amounts of gain affect signal levels:

- $|A| > 1$: At high gains, the inverter pushes the output signals closer to the power supply rails. When $V_{in} = V_{IL}$, the output voltage magnitude will be greater than V_{IL}.
- $|A| < 1$: At gains less than unity, an input voltage $V_{in} = V_{IL}$ will produce an output voltage magnitude that does not reach V_{IL}. The inverter's output voltage is closer to the unknown range than was its input voltage.

High gains allow us to **restore** logic levels, an important property for reliability.

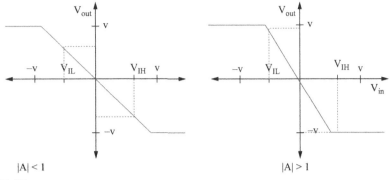

FIGURE 4.7

Effects of high and low gain on signal levels.

4.2.4 **Gain and delay**

Gain also improves the delay through logic gates—higher gain results in shorter delay. These shorter delays come at the cost of increased energy consumption, but choosing the gain of a logic gate is an important design optimization.

Delay is a dynamic property of the circuit; the saturating amplifier model gives us a simple tool for understanding delay in cascaded logic. To understand the importance of gain to delay, we will use as an input a ramp voltage that steadily increases from V_{SS} to V_{DD} starting at time t. As we saw in Chapter 3, the ramp is a more realistic model of an input signal than was the step input we used to formulate the RC delay model.

Fig. 4.8 shows the response of the inverter for different values of A. In the first case, $|A| = 1$, so the output voltage is equal to the negative of the input voltage. This means that time at which the inverter's output changes from 1 to X is the same as the time required for the input to change from 0 to X. The unity gain of the inverter means that it has maintained the delay contained in the ramp signal.

When we use our saturating inverter model, we can write the delay through a series of n inverters as

$$V_{out}(n) = (AV_{in}(0))^n \qquad (4.6)$$

When the inverter's gain is greater than 1, the output waveform has a steeper slope than the input. This means that the time of the $1 \rightarrow X$ transition at the output is less than the time required for the input to transition from 0 to X. If the inverter's gain is less than 1, then the output transition takes longer than the input transition. In this case, the gate has slowed down the signal.

The situation is worse than Eq. (4.6) would suggest—gain depends on input slope so slow-rising inputs reduce the gain of the gate and increase its delay. Our assumption

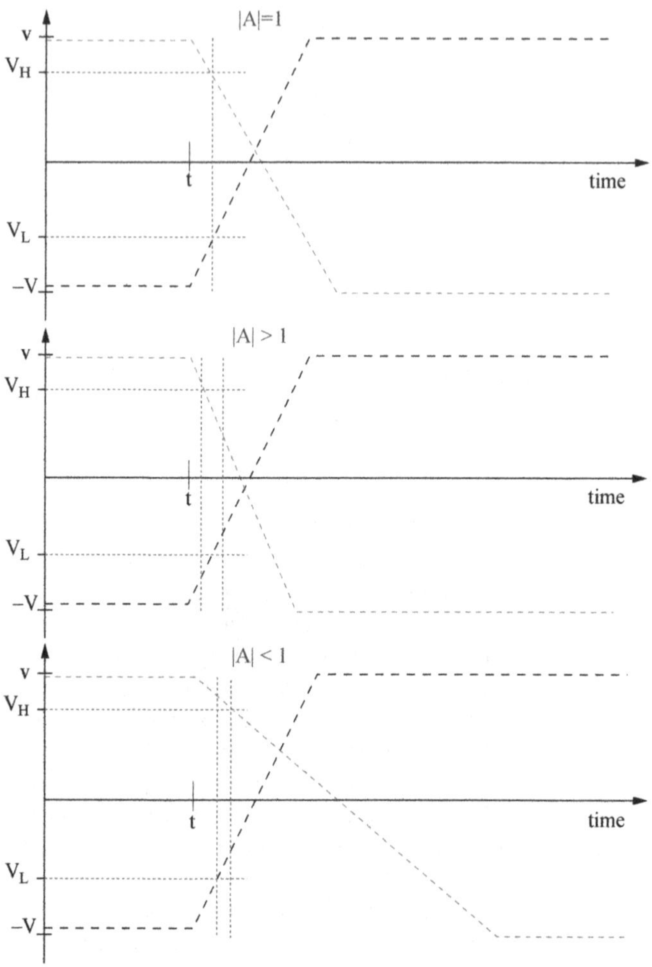

FIGURE 4.8

Ramp response of an inverter.

of an ideal step function simplified our analysis—an ideal step connected to the gates of the two transistors would cause one to turn on and the other to turn off instantly. However, real gates do not receive ideal waveforms, as shown in Fig. 4.9. The input to a gate comes from the output of another gate, so realistic inputs take time to transition from one logic level to the other.

Transistor behavior in gates

The result of a ramping input voltage is to cause the transistors to turn on and off more slowly, reducing the amount of current available to drive the output. Consider the case of a rising input to an inverter, which causes its n-type transistor to turn

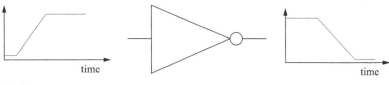

FIGURE 4.9

Nonideal waveforms.

on. Fig. 4.10 shows two cases: a fast-rising input and a slow-rising input. In the first case, the transistor's gate voltage will rise quickly with the input. As it does so, the drain current increases quickly to its maximum possible value, the saturation current at $V_{gs} = V_{DD}$. As the capacitance at the gate's output starts to drain, V_{ds} decreases, causing the drain current to drop. The transistor's ability to drain the load capacitance is greatest when it is in saturation and its gate voltage is at its highest value. When the input rises slowly, the gate voltage also rises slowly. In this case, the transistor is in saturation and drains current from the load, but it does so at a lower rate than if the gate voltage were V_{DD}. As a result, the drain voltage drops significantly as the input voltage ramps up. When the input voltage reaches V_{DD}, the drain current is closer to the linear region boundary. This means that the transistor has spent less time at its maximum current, slowing down the transition.

Fig. 4.11 compares the response of gates to step and ramp inputs. The output voltage initially stays high because the input voltage has not yet risen high enough to turn on the pulldown. Once the pulldown's gate does reach its threshold voltage, its drain current slowly increases, causing the shoulder of the response.

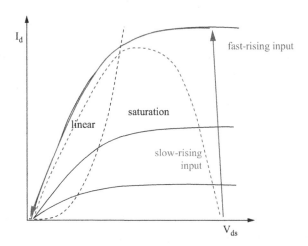

FIGURE 4.10

Trajectory of transistor current during fast- and slow-rising inputs.

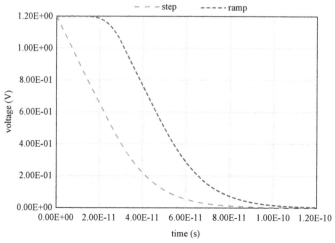

FIGURE 4.11

Voltage waveform for step and ramp inputs.

Many other logic families, particularly those built with other types of transistors, do not exhibit this slope-dependent delay to nearly the same degree. MOS transistors do not have large gains compared to, for example, bipolar transistors. Logic gates based on bipolar transistors, such as TTL, have enough gain that their output slope is independent of the input slope. The relatively low gain of CMOS gates requires us to pay extra attention to the gain of gates and the sizes of their transistors.

4.2.5 Delay and power

As we saw in Chapter 3, increasing the sizes of transistors to decrease delay costs energy. However, when we are designing complex logic networks, only some of the gates matter. Gates that are not on the critical path are faster than they need to be. These gates can be slowed down, saving energy.

Consider the logic network of Fig. 4.12. If all the gates have $\delta = 1$, then g4 is not on the critical path. If we slow down g4 to $\delta = 3$, perhaps by making its transistors less

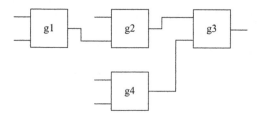

FIGURE 4.12

Effects of slowing down gates on the critical path.

wide to consume less power, it will join the critical path. At this point, we cannot slow it down further without slowing down the entire network.

4.2.6 Noise and reliability in logic and interconnect

Signal integrity refers to the challenges posed by ensuring that wires carry correct, uncorrupted values. We moved to digital circuits in large part to minimize the effect of effects that corrupt signals in analog circuits. But digital circuits are also susceptible to noise. As the sizes of transistors and wires scale with Moore's Law, signal integrity becomes increasingly challenging. Fig. 4.13 outlines the sources of noise in combinational logic [She98]:

- Variations in power and ground voltages affect the operation of gates as we will see in the next section.
- Crosstalk between wires introduces noise. We will study this problem in detail in Section 4.3.
- Coupling of a gate's output to its input affects its delay as will be shown in Section 4.2.8.

4.2.7 Power supply and reliability

We have assumed that the power supply signals, V_{DD} and V_{SS}, are reliable and constant values. That assumption is not valid in realistic systems. V_{DD} and V_{SS} are electrical signals just like any other and equally susceptible to electrical problems.

One source of variations in power supply voltage is due to the resistance of the wires used to carry power. The current that flows through the wires causes a voltage drop that results in the gates seeing a lower power supply voltage than the one presented at the power supply terminals.

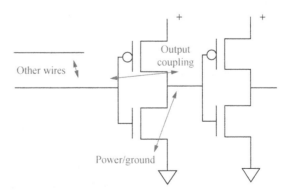

FIGURE 4.13

Sources of noise in digital circuits.

Voltage drops across power supply wiring occur for two major reasons: resistance and inductance. We will see that one important solution to wiring-driven voltage fluctuations is to use capacitance.

Example 4.1 Metal Wire Resistance

A minimum-width wire on the first level of metal in our example process has a resistance per unit length of 0.44 Ω/μm. If our transistor delivers a current of 150 μA, then we can find the length of the wire that generates a 0.1 V drop as

$$0.1 \text{ V} = 150 \text{ μA} \times 0.44 \frac{\Omega}{\text{μm}} \times l \text{ μm}$$

Solving for l,

$$l = 1515 \text{ μm}$$

We can fight the resistance problem by increasing the width of the power supply wires. The width of the wires required depends on the number of gates fed by the wire and the current those gates require. As shown in Fig. 4.14, the power and ground nets are organized as a pair of trees, with the branches of the two trees interdigitated. We use relatively narrow wires connected to the gates, each wire connected to only a few gates. We feed each of those wires with larger wires. We combine V_{DD} and V_{SS} wires in a tree, with each level of the tree requiring wider wires.

Modern chips use several metal layers for power distribution: lower-level layers provide local distribution to gates while higher layers distribute power globally. Wires on higher interconnect layers are thicker, giving them lower resistance and making them more suitable to carry large currents.

Pin inductance   The pins on the package that contains the chip have parasitic inductance that can cause transient voltage drops during chip operation. Large chips consume significant amounts of current. That large current creates voltage drops across the pins that connect the chip to the printed circuit board. The inductance of those pins is

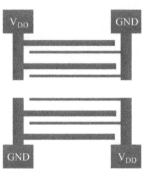

FIGURE 4.14

The structure of an on-chip power supply network.

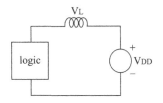

FIGURE 4.15

Inductive voltage bounce in logic.

nonnegligible. The voltage drop across an inductance depends on dI/dt. The current drawn by gates varies over time as they switch; as we aggregate the current from many gates together at the power supply pins, the currents can become large and their derivatives can also become large.

As illustrated in Fig. 4.15, rapidly changing currents on the power supply lines result in large inductive voltage drops across the pins. The voltage drop across the power supply pins of a high performance chip can become very large. We can reduce the voltage swing caused by current fluctuations is to spread the power over multiple pins. This technique reduces the amount of current per pin and therefore dI/dt.

Example 4.2 Processor Pinout

The Intel Xeon Processor E7-8800 [Int11] has a total of 1581 pins. Of these, 381 supply power and 471 are ground connections.

A typical value for the combined inductance of a pin and the circuit board wire to which it is attached is 0.5 nH. If we assume that each pin sees a maximum current fluctuation in each of the 471 ground pins of

$$\frac{dI}{dt} = \frac{0.1 \text{ A}}{10^{-8} \text{ s}} = 10^7 \frac{A}{s}$$

then the voltage across the pin is

$$V_L = L\frac{dI}{dt} = 0.005 \text{ V}$$

If the current fluctuation is concentrated on a single pin, then the maximum voltage drop is

$$V_L = L\frac{dI}{dt} = (0.5 \text{ nH})\left(4.7 \times 10^9\right) = 2.3 \text{ V}$$

That voltage drop is considerably larger than the power supply voltage itself.

Circuit effects   Variations in power supply voltage can occur at either terminal. For simplicity, we will consider variations in the ground signal, a condition often referred to as **ground bounce**; the V_{DD} signal is also subject to bouncing. As shown in Fig. 4.16, we can model ground variations as a voltage supply between ground and the gate. If every gate saw the same ground bounce, the gates would simply become slower.

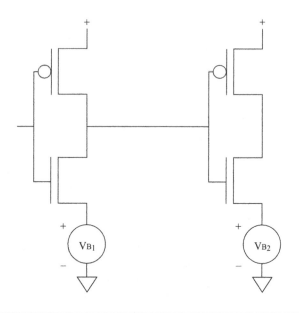

FIGURE 4.16

Circuit model for ground bounce and logic gates.

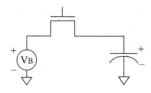

FIGURE 4.17

Circuit model for ground bounce and current.

The circuit of Fig. 4.17 shows more concisely the situation in which a pulldown transistor in a gate experiences ground bounce. The transistor is on and drawing current from the capacitor to the ground to create a $1 \rightarrow 0$ transition. A positive ground bounce voltage reduces the driving transistor's V_{ds}, thus reducing current drive. Since the current depends on V_{ds}^2, this effect is particularly important and is larger than any reduction in the output voltage caused by operating at a reduced voltage.

The news becomes even worse. If series-connected gates see different amounts of ground bounce, the logic values transmitted between them become unreliable. As shown in Fig. 4.18, the ground bounce voltage changes the gate's transfer characteristic. The output low logic threshold moves significantly; the final resting voltage of the gate for a low output is now above V_{SS}.

Fig. 4.19 shows a test circuit for different amounts of ground bounce at different gates. The first gate has seen an elevated ground voltage while the second gate stays at

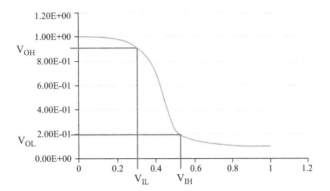

FIGURE 4.18

The effect of ground bounce on the gate transfer characteristic.

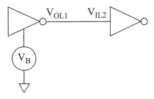

FIGURE 4.19

Effect of ground bounce on gate-to-gate communication.

V_{SS}. The logic 0 output voltage of the first gate V_{OL1} does not go all the way to V_{SS}. If the ground bounce is large enough, then $V_{OL1} > V_{IL2}$ in which case the first gate thinks it has sent a logic 0 but the second gate sees the value as an X. Even if the bounce is not large enough to send the signal all the way to X, the designer's noise margins have been reduced and the signal can be corrupted by the addition of other sources of noise.

The current demands of various regions of the chip can vary in both time and space. As shown in Fig. 4.20, a large chip has many blocks of logic, each performing its own function. The activity of each block may change over time—it may have many changing inputs at one time and very few at another. Different blocks will often have different activity patterns. As a result, the current demands follow complex patterns. Even though the average current draw is reasonable, the power distribution network has to be designed to handle their maximum loads.

One way to reduce power supply fluctuations is to add capacitance between the power supply wires. This capacitance is known as **decoupling capacitance** C_D since it decouples the gates at the load from the power supply. As shown in Fig. 4.21, the decoupling capacitor provides a reservoir of charge that can be used to supply the load current if the power supply wires cannot supply enough current on their own.

We typically specify the maximum voltage droop ΔV allowed across a gate. A decoupling capacitor will be connected across a set of n logic gates. If we assume

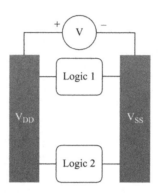

FIGURE 4.20

Logic blocks and current demands.

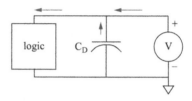

FIGURE 4.21

A decoupling capacitor for the power supply wiring.

that each gate draws I_{max} current for time t_{max}, then we can estimate the charge required to supply the gates during the surge as

$$Q_{max} = nI_{max}t_{max}. \tag{4.7}$$

For simplicity, we will assume that all the charge during the current surge is supplied by the decoupling capacitor. We need enough decoupling capacitance C_D to supply the surge charge:

$$C_D = \frac{nI_{max}t_{max}}{\Delta V} \tag{4.8}$$

Example 4.3 Decoupling Capacitance

We need to provide a coupling capacitance for $V = 1$ V, $\Delta V = 0.1$ V for $n = 10$ gates for 1 ps at a maximum current of 150 μA. Then

$$C_D = \frac{10 \times 150\ \mu A \times 1\ ps}{0.1\ V} = 15\ fF$$

Highlight 4.1

V_{DD} and V_{SS} are subject to noise.

4.2.8 **Noise and input/output coupling**

One interesting source of noise that we have not yet analyzed is input/output coupling. The gate-to-substrate capacitance is not the only capacitance at the transistor's gate. The gate material also has some capacitance to the source and drain of the transistor thanks to their close proximity. While the gate-source and gate-drain capacitances C_{gs} and C_{gd}, illustrated in Fig. 4.22, are smaller than the gate-substrate capacitance, they are large enough to cause problems, thanks to the amplification of the inverter.

As shown in Fig. 4.23, C_{gs} and C_{gd} are connected between the input and output of the inverter. This coupling slows down the inverter thanks to its inverting amplification. If the input value falls, causing the output to rise, the input/output capacitance conducts some current from the output back to the input. The feedback current fights the input current, causing the inverter's output to change more slowly.

The **Miller effect** [Mil20] allows us to approximate this effect very simply. As shown in Fig. 4.24, the input/output coupling capacitance C is

$$C_M = C(1 + A) \tag{4.9}$$

where A is the gain of the inverter.

In the case of input/output coupling in a logic gate, a relatively small C_{ds} can cause a significant increase in the capacitive load of a gate. The effect of C_{ds} coupling is

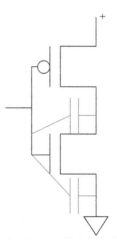

FIGURE 4.22

Coupling between the input and output of an inverter.

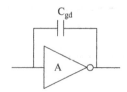

FIGURE 4.23

Miller capacitance.

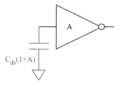

FIGURE 4.24

Gate/substrate and Miller capacitances.

even worse for gates with more than one input, causing effects similar to crosstalk on wires. If one input causes the gate's output to change, the input/output capacitances will feed back values to all the inputs. As a result, even signals that are not being driven to change can see disturbances.

One interpretation of the Miller effect is that signals flow backward through the logic. Our event model of Section 4.2.1 assumed that events flowed from gate inputs to gate outputs. The input/output capacitances on the gates are, however, bidirectional, and they allow current to flow both ways.

4.3 Interconnect

In this section, we look at the properties of wires, which have their own physical properties that limit their electrical performance. Wiring characteristics—resistance, capacitance, inductance—create some of the most fundamental limits on chip performance. The traditional assumption made in circuit design—that a wire is a point object with the same voltage and current everywhere—is invalid in VLSI systems.

4.3.1 Parasitic impedance

Wires, being physical objects, are not ideal conductors. When designing chips, we cannot ignore those physical characteristics. On-chip wiring exhibits resistance, capacitance, and inductance. We refer to those values as **parasitic** because they are secondary—we did not design the wires specifically to have those properties.

On-chip wiring is designed to precise vertical dimensions. While different layers of wiring may have different thicknesses, the thickness of all the wires at any given level is uniform. This allows us to discuss on-chip wires in units of area for most purposes. Circuit designers have control over the length and width of wiring material, just as they control the width of a transistor channel.

Because the height of on-chip wires is fixed by the manufacturing process, wire characteristics are usually not quoted as resistivity but rather as **sheet resistance** in units of $\Omega/\blacksquare$ (ohms per square). We can understand the appropriateness of ohms per square as a measurement using either of two lines of reasoning. First, the resistance of a piece of material with width W, length L, and height H is $R = \rho L/(HW)$; if we double the length and width, leaving the height constant, the resistance is $R = \rho 2L/[H(2W)]$. Second, consider the squares of material shown in Fig. 4.25 with each small square having a resistance of 1 ohm. If we form a 2×2 array of squares of material, they combine in series along the direction of current flow and in parallel perpendicular to current flow. As a result, the larger square has the same resistance as each of its constituent squares—1 ohm.

To measure the resistance of a nonsquare piece of material, we must know the direction of current flow, as shown in Fig. 4.26. In the example, the current must flow through three squares of material in series; if current were to flow in the perpendicular direction, the squares would be in parallel. Corners of wires have more complex current flows; we generally approximate a corner between two equal-width wires as having a resistance of $^{1}/_{2}\,\blacksquare$.

Wire capacitance for a manufacturing process is typically given as a **unit capacitance** or capacitance per unit area. We measure unit capacitance per unit area in units such as $fF/\mu m^2$. The thickness of the capacitor oxide is determined the manufacturing process much as is wire thickness. Circuit designers control

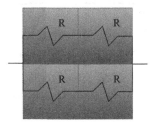

FIGURE 4.25

Resistance of squares is independent of square dimensions.

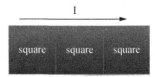

FIGURE 4.26

Wire resistance is measured along the direction of current flow.

the area of the plates and use the unit capacitance value to determine the capacitance of their structure.

A more detailed model includes both the parallel plate capacitance as well as **fringing capacitance**. The electric field at the edges of the plate is distorted as compared to the perpendicular field of infinitely large parallel plates, which changes the capacitance at the edge. A more accurate measure of capacitance would combine the parallel plate value with an additional term that depends on the length of the capacitor's perimeter. The pn junctions also contribute **junction capacitance**. Since a diffusion wire is embedded in a tub of the opposite carrier concentration, the junction capacitance is measured along the entirety of the wire. Junction capacitance is typically measured separately for the bottom and sides of the diffusion wires.

Example 4.4 Resistance and Capacitance Measurements

Consider this piece of metal:

The metal wire is 0.1 μm wide and 2 μm long. If the resistivity of the metal is 0.04Ω/■, then the resistance of the wire as current flows along its length is

$$R = R_\blacksquare \frac{L}{W} = (0.04\Omega/\blacksquare)\frac{2\ \mu m}{0.1\ \mu m} = 0.08\ \Omega$$

We can also calculate the capacitance of this wire to the substrate. If the unit capacitance area is 50 aF/μm², then the total wire capacitance is

$$C = C_A WL = \left(50\ aF/\mu m^2\right)(0.1\ \mu m)(2\ \mu m) = 10\ aF$$

4.3.2 Transmission lines

Long wires are critical to modern computer design, and we need to understand their properties. We have so far treated parasitic values as **lumped**—a single component. For example, ideal scaling treated wire resistance as lumped. However, under many conditions, the lumped assumption is not accurate. In these cases, we must model the wire using a **distributed** model called a **transmission line**. Delays through transmission lines are greater than through lumped wires—the lumped assumption for wire delay made by Dennard et al. was, in fact, optimistic.

Fig. 4.27 shows a simple transmission line. The transmission line itself is formed by a pair of conductors. We drive the transmission line in this case with a voltage source V_S; we also model the impedance at the source of the wire as Z_S. At the other end of the transmission line, a load Z_L connects the two conductors, forming a circuit. The bottom conductor forms the return path for current.

We measure voltage or current in a lumped element as a single value—an ideal, lumped resistor has a single voltage across it, for example. That lumped voltage or

FIGURE 4.27

A transmission line.

current is a function of time. A transmission line has physical size; we must measure both the voltage across the conductors and the current through them as a function of *position* as well as of time.

We can model the transmission line as a series of **sections**. Each section has lumped elements for resistance, capacitance, and inductance. Each section's voltage and current is a function of time; the variations from section to section model dependencies on position.

Telegrapher's equations   The general model for the behavior of a transmission line with resistance, inductance, and capacitance is known as the **telegrapher's equations** since the first use of electrical transmission lines was for telegraphy. A single section of a transmission line includes series and parallel elements as shown in Fig. 4.28; the G element is a conductance.

We can write the voltage and current of a general transmission line as partial differential equations:

$$\frac{dV(x)}{dx} = -(R + j\omega L)I(x) \tag{4.10}$$

$$\frac{dI(x)}{dx} = -(G + j\omega C)V(x) \tag{4.11}$$

Large-scale transmission lines such as those used for telecommunications are often modeled as having negligible resistance. These lines are known as LC transmission lines. On-chip wires, due to their small size, have significantly different properties. We usually cannot neglect the resistance of on-chip wires. In at least some cases, however, we can ignore their inductance due to two factors: high resistance values mean that inductive effects are not significant at the frequencies of interest; and the wires are not long enough to show significant inductive effects. Detailed analysis of on-chip interconnect is based on an RLC transmission line model; for our purposes we will concentrate on a simpler but still useful model.

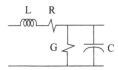

FIGURE 4.28

A section of an RLC transmission line.

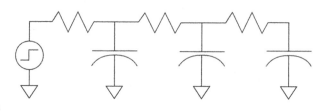

FIGURE 4.29

An RC transmission line.

RC transmission lines are of practical use in chip design; they are also much simpler to understand than RLC transmission lines. Fig. 4.29 shows a model for an RC transmission line. The section resistance is in series while the capacitance is in parallel with the signal to model the interactions between the two conductors. The line is driven by a step input.

To understand how signals are propagated by transmission lines, assume that the capacitors start out at 0 V before the step is applied. Fig. 4.30 shows how the signals propagate over time and along the length of the wire. Just after the step is applied, the

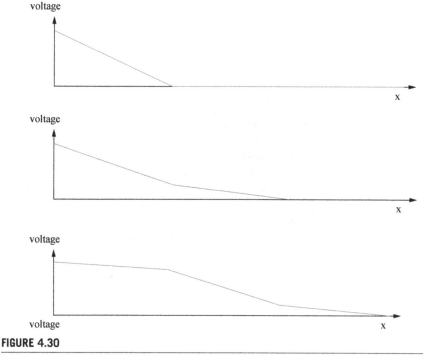

FIGURE 4.30

Propagation of a step input along an RC transmission line.

entire step voltage appears across the first section's resistor since the capacitor is at 0 V and $V_{in} + V_{R1} + V_{C1} = 0$.

The current through that resistance starts to charge the first section's capacitance. As the first section charges, it applies a voltage across the second section. The current through the second section's resistance will initially be small since its voltage is small: $I_{R1} = I_{C1} + I_{R2}$.

But as the first section charges, the current flowing to the second section will increase. The effect will repeat itself at the next section. As a result, the square wave propagates along the wire. The resistance of the section means that the original pulse is **distorted**. As the pulse travels along the wire, its slope decreases.

If we apply a pulse to the transmission line, the effects of the transmission line become even clearer, as shown in Fig. 4.31. The pulse travels along the wire at a finite

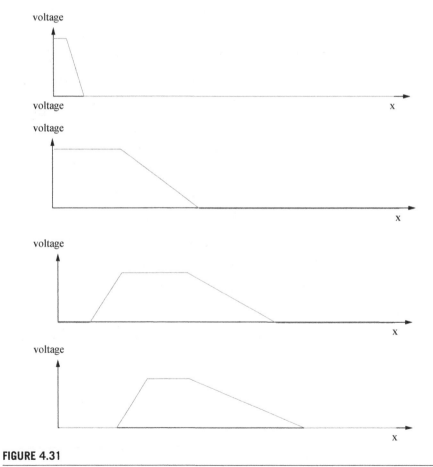

FIGURE 4.31

Distortion of a pulse traveling along a transmission line.

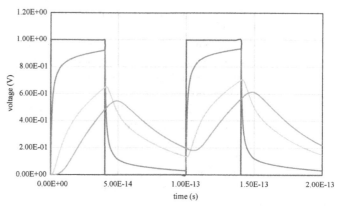

FIGURE 4.32

Performance of a series of RC sections.

speed—the wire itself introduces delay. Furthermore, both the leading and trailing edges of the pulse are distorted, and its magnitude is attenuated.

Fig. 4.32 shows the results of a circuit simulation of the response of an RC transmission line when driven by an ideal voltage source. The plot shows waveforms at several points along the line: the input pulse, the first RC section output, the fifth section output, and the tenth section output. Each section has $R = 10\,\Omega$, $C = 100$ aF. Even the first pulse has been distorted by the load presented by the rest of the transmission line—with a time constant of 1 fs, the first section should reach near its asymptotic value after 5 fs but has failed to do so after 40 fs. The pulse is even more distorted and delayed at later stages.

Elmore delay

The **Elmore model** [Elm48; Rub83] gives an elegant estimate of delay through an RC transmission line. The resistance and capacitance of a section are r and c, respectively; we do not require all the sections to have the same resistance and capacitance. The **Elmore delay** of the wire is

$$\delta_E = \sum_{1 \le i \le n} c_i \sum_{1 \le j \le i} r_j \tag{4.12}$$

As shown in Fig. 4.33, each capacitance is charged through the resistance of all the sections leading up to it; for example, c_2 is charged through r_1 and r_2.

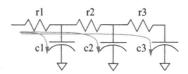

FIGURE 4.33

Currents in the Elmore model.

If all the sections have the same resistance and capacitance, the Elmore delay has a particularly simple form:

$$\delta_E = \sum_{1 \le i \le n} (n-i)rc = \frac{1}{2}rcn(n+1) \qquad (4.13)$$

This means that delay through a uniform wire is proportional to the *square* of its length. This result is much more pessimistic than the linear dependency assumed by ideal scaling that we saw in Section 3.6.

Highlight 4.2

Elmore delay is proportional to the square of wire length for uniform-width wires.

Example 4.5 Elmore Delay of a Uniform Width Wire

Consider our metal wire from above with dimensions of 0.1 μm × 2 μm. If we divide it into $n = 10$ sections, then the resistance of each section is

$$r = R_\blacksquare \frac{L}{W} = (0.04 \ \Omega/\blacksquare)\frac{0.2 \ \mu m}{0.1 \ \mu m} = 0.08 \ \Omega$$

And the capacitance of a section is

$$c = C_A WL = \left(50 \ aF\big/\mu m^2\right)(0.01 \ \mu m)(2 \ \mu m) = 1 \ aF$$

The Elmore delay through this wire is

$$\delta_{rc} = \frac{1}{2}(0.08 \ \Omega)(1 \ aF)(10)(11) = 4.4 \times 10^{-18} \ s$$

Now consider a wire of length 2 mm or 1000X longer than our original wire. We can use the same section size but increase n to 10,000. In this case,

$$\delta_{rc} = \frac{1}{2}(0.08 \ \Omega)(1 \ aF)(10,000)(10,001) = 4.0 \times 10^{-12} \ s$$

Nonuniform-width wires are also useful. Increasing the width of a wire along its entire length would reduce its resistance but also increase its capacitance. However, making the wire wider near its source and narrower near its sink exploits the properties of Elmore delay. The resistance at the wire's source is applied to charging the capacitance along the wire's entire length as illustrated in Fig. 4.34. Making the wire wider at its source reduces its capacitance, improving the delay of all the sections. A thin wire at the far end has more resistance but also less capacitance. It can be shown that an optimally shaped wire is exponentially tapered, another example of impedance matching [Fis95].

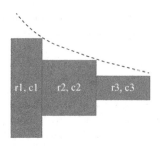

FIGURE 4.34

Resistance and capacitance along a nonuniform wire.

4.3.3 Crosstalk

Unfortunately, transmission lines cannot always be considered in isolation. If two transmission lines travel side-by-side for a long distance, the parasitic capacitance between them is significant. As shown in Fig. 4.35, a wire has capacitance to any other adjacent conductor. A wire that is directly above the substrate will have a capacitance to that substrate. The substrate is connected to the power supply; the capacitance slows down transitions but is not a source of noise (ignoring power supply noise). But the wire is also capacitively coupled to the wires above it; at higher levels of the interconnect hierarchy, the wire will not have significant capacitance to the substrate but will be coupled to the wiring layers above and below. A wire is also coupled to horizontally adjacent wires; in this case, the parallel plates of the capacitor are formed by the vertical walls of the wires. All these coupling capacitances connect wires each of which has its own signals.

Wire-to-wire coupling causes **crosstalk**—the transitions on one wire are transmitted to the other wire. If the receiving wire's signal is stable, crosstalk can induce a transition on the wire; if the receiving wire's signal is moving in the opposite direction, crosstalk fights and delays the intended transition.

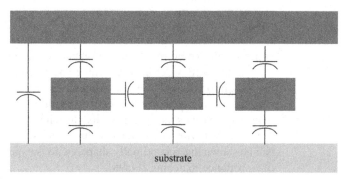

FIGURE 4.35

Capacitive couplings of wires.

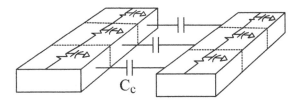

FIGURE 4.36

Capacitive coupling between wires.

As shown in Fig. 4.36, we can analyze crosstalk by considering a transition on the **aggressor** wire that causes crosstalk on the **victim** wire. The coupling capacitance per wire section is C_C, and the substrate capacitance per section is C. The amount of charge delivered by the aggressor pulse to the victim in time Δt is

$$\Delta q = I_C \Delta t = C_C \frac{\Delta V_A}{\Delta t} \Delta t = C_C \Delta V_A \qquad (4.14)$$

This charge is then shared between the coupling capacitance and the victim's substrate capacitance. This causes a voltage change in the victim:

$$\Delta V_V = \frac{C_C}{C + C_C} \Delta V_A \qquad (4.15)$$

On a typical chip, the coupling capacitance C_C is two to four times larger than the substrate capacitance C. Since the coupling capacitance is larger, crosstalk is a very significant effect in modern chips.

Several techniques can be used to minimize the effects of crosstalk. Fig. 4.37 shows signal wires interleaved with ground wires. The same amount of coupling exists between each signal wire and its adjacent ground lines as would be the case with

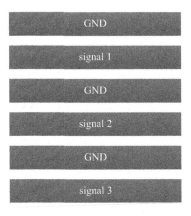

FIGURE 4.37

Shielded wires.

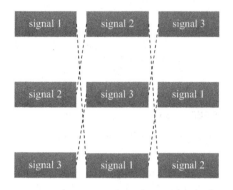

FIGURE 4.38

Twizzled wires.

signal wires, but the ground signal is stable. Because ΔV is smaller, less crosstalk is injected into the signal lines. Fig. 4.38 shows a technique known as **twizzling**. Rather than run each signal along straight lines, the wires jump from track to track periodically. The vertical connections between signal sections are made by vertical metal wires and vias that are represented by the dotted lines for clarity. Once again, the same amount of coupling capacitance exists. But in this case, if the signals are uncorrelated, less charge is injected into each signal on average.

4.3.4 Wiring complexity and Rent's Rule

Different logical functions require different wiring. Some functions require many short connections while others require more long connections. We can estimate the complexity of the wiring required for a function using partitioning. Consider the logic of Fig. 4.39 which has been partitioned into two pieces. If we can find the partition

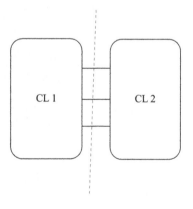

FIGURE 4.39

Partitioning as a measure of wiring complexity.

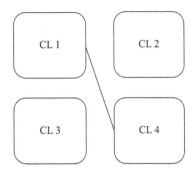

FIGURE 4.40

Recursive partitioning for wiring estimation.

that gives us the smallest number of wires that cross the boundary between the two, then we have an estimate of the number of wires that have to go long distances across the function.

As shown in Fig. 4.40, we can recursively partition the logic to provide a more fine-grained estimate of wiring complexity. This recursive partitioning can be used to guide the **placement** of components. If we recursively subdivide the regions of the chip, we can map the logic partitions onto those regions. Partitioning will group together gates that are closely connected.

Wire length estimation   We can estimate wire length given a placement. Wire length helps us to estimate which wires will be on the critical timing path; we can adjust the placement to shorten the critical path wires and reduce delay. Fig. 4.41 shows a placement of gates. Wires between the gates have been drawn as straight lines. The wires on the masks must be formed by rectilinear segments, but Euclidean distance is a simple estimate of wire length. We can change wire lengths by swapping components in the placement. But since the gates are connected in a number of ways, several wires may change length.

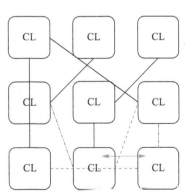

FIGURE 4.41

Wire length estimation by Euclidean distance.

If, for example, we swap the two components at the bottom right, two wires will become shorter while two will become longer.

Rent's Rule   **Rent's Rule** [Lan71] allows us to estimate the number of pins required by a function even without partitioning. The rule is based on measurements of a number of logic designs. The parameters vary somewhat based on applications, but Rent's Rule has been verified to hold for many different technologies. The rule states that the relationship between pins N_p and components N_g is log/log:

$$N_p = K_p N_g^r \tag{4.16}$$

where r is known as **Rent's constant**. Landman and Russo [Lan71] used parameter values in the ranges $0.57 \leq r \leq 0.75$, $1.5 \leq K_p \leq 2$; some others use $K_p = 2.5$. Others have found the values $r = 0.45$, $K_p = 0.82$ to be a better fit for modern microprocessors; the smaller number of blocks at high levels of abstraction changes the relationship between logic and pinout.

4.4 Sequential machines

Our logic model is both too simple and too complex. On the one hand, we can compute only relatively simple functions using only combinational logic. On the other hand, the event model requires us to keep track of a great deal of event activity in the logic to determine how the outputs change. We will first develop models for sequential machines. We will then consider clock period as a metric. Section 4.4.4 analyzes metastability, an important failure mode for sequential machines.

4.4.1 Sequential models

An intermediate step is to model logical functions as **combinational logic** rather than collections of gates. As shown in Fig. 4.42, a combinational model block is described by the Boolean function it performs.

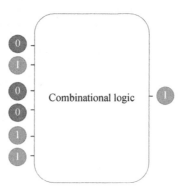

FIGURE 4.42

The combinational abstraction for logic.

The details of the gates used to perform the function affect the delay and energy consumption of the combinational logic—we can build different combinational logic blocks that perform the same function but at different costs in delay and energy. However, we can abstract the delay to the worst case from input to output, as we saw with critical path analysis.

The combinational logic abstraction helps us to model time as discrete. When we analyzed logic delay in Chapter 3, time was real-valued. The event model of Section 4.2.1 was based on gate delays, so its model of time was also real-valued. As shown in Fig. 4.43, time can be represented by integers, and signals take on values only at discrete, integer-valued times: $<0,1,2,\ldots>$.

However, the combinational logic abstraction is not enough. The Turing machine assumes discrete time. The **sequential machine** (or **state machine**) model provides us with a way to build machines that effectively operate in discrete time. As shown in Fig. 4.44, the machine has a **state**, stored in a register. The behavior of the machine is described by its sequence of states. The register's input is D and its input is Q. It is controlled by a clock signal, sometimes called φ, which determines when the register reads its D input, stores the value, and presents that stored value at the D output.

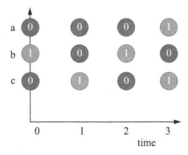

FIGURE 4.43

Signals in discrete time.

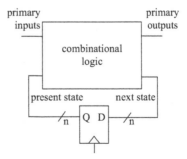

FIGURE 4.44

A sequential machine.

The machine's behavior is defined by two functions that are performed by the combinational logic: its **state transition** function maps the primary inputs and state to the next state; the **output function** maps the primary inputs and state to the primary outputs. Fig. 4.45 shows an example state transition table and state transition graph, two equivalent ways of enumerating the FSM's state transition and output functions. The state transition table is a Boolean truth table that gives the state transition and output functions. Some sequential machines are not naturally described in this form. For example, a multiplier connected to a register would not be easy to describe as a state transition table. Although we may use different notations to describe sequential behavior, any sequential machine can be described in these forms.

Any problem that causes the FSM's registers to record an incorrect state value creates a **permanent fault**. That value will be recirculated repeatedly, causing continued

State transition table

input	present state	next state	output
0	00	00	0
1	00	01	0
0	01	01	0
1	01	10	0
0	10	10	0
1	10	00	1

State transition graph

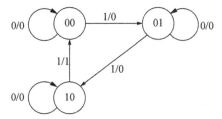

FIGURE 4.45

State transition tables and graphs.

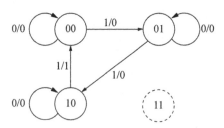

FIGURE 4.46

An FSM with an unspecified state.

errors in the machine's operation. Fig. 4.46 shows the state transition graph of Figure comb-stg and its unspecified extra state. The state bit value 11 has not been specified in the machine. Exactly what the machine does if the registers are corrupted to that state depends on the logic functions implemented for the next state and output. A register can be set to a bad value either by errors in the combinational logic that load a bad value into the registers or by corrupting the value in the register itself.

4.4.2 **Registers**

To build a sequential machine, we need a circuit that can be used to store a value. We will use the term **register** for any circuit that stores a single bit. Several different types of circuits are used to build registers, each with its own advantages and disadvantages. We can define two different axes along which to categorize registers:

- How they store their value.
- How they react to the clock signal.

Static and dynamic registers

Registers can use one of two different storage mechanisms:

- A **dynamic register** uses the gate capacitances on an inverter to store a value.
- A **static register** uses feedback between a set of gates to store the value.

Each technique has its own advantages and disadvantages. Fig. 4.47 shows the circuit schematic for a dynamic latch, which consists of an inverter guarded by an input transistor. The value is stored on the gate capacitance of the inverter. When the access transistor is off, the output Q' is equal to the inverse of the value stored on the gate capacitance. (The prime symbol denotes logical inversion.) When $\varphi = 1$ and the guard transistor is on, the input D can be used to charge or discharge the gate capacitance, changing the value of the register. The standard symbol for the clock signal is φ.

Fig. 4.48 shows a schematic for a static register. The value is stored in the connection between the two inverters. When the clock signal is low, the access transistor is off and the transistor in the feedback path is on. The stored value is fed back between the inverters to reinforce and restore its value; so long as power is applied to the register, the value will remain. When the clock goes high, the access transistor turns on

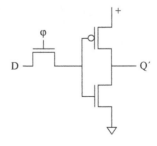

FIGURE 4.47

Circuit schematic for a dynamic latch.

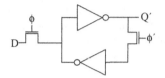

FIGURE 4.48

A static latch circuit.

and the feedback transistor turns off, allowing the external input to override the latch's value. Static registers are subject to corruption from the noise effects we discussed in Section 4.2 [She98]. The amplification of the feedback gates that store the value helps to maintain the proper value.

Dynamic registers are small and consume little power. However, the value can be corrupted by leakage of charge from the gate capacitance. Charge in capacitors is subject to leakage over time. When enough charge leaks away, a 1 degrades to an X or 0. The charge on a dynamic latch is valid for about 1 ms at room temperature. If the register is clocked fast enough, its value will be refreshed and leakage does not matter. But some latches are left unclocked for extended periods, particularly in sleep modes. Registers that may go without clock signals for long periods should use static circuits.

Latches and flip-flops

The other way in which we classify registers depends on how they react to the clock signal:

- **Latches** are *transparent* and *level-sensitive*.
- **Flip-flops** are not transparent and are *edge-triggered*.

A latch is transparent because, when the clock is high, its output will change as it follows the input. When the clock input to the dynamic latch of Fig. 4.47 is high, the access transistor creates a circuit path from the D input to the Q' output. As a result, the latch's output will change in step with its input.

We say that a flip-flop is edge-triggered because it stores the new value at a clock transition. Fig. 4.49 shows one type of flip-flop, the master-slave, which is composed of two latches. (The flip-flop can be either static or dynamic depending on which type of latch is used.) The first latch receives an uninverted version of the clock while the second latch receives an inverted clock. When the clock is high, the first latch is transparent and reads its input value. The second latch is closed and retains its old value. When the clock signal goes low, the first latch closes and the second one opens. The flip-flop state changes only at the clock edge.

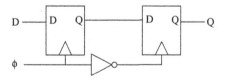

FIGURE 4.49

A master-slave flip-flop.

Latches require more complex clocking than do flip-flops. Latches must be organized into multiple ranks, each clocked by a different **clock phase**. The clock phases ensure that no cycles are formed that would cause the machine to oscillate.

Register timing

Proper operation of a register requires that some relationships between the timing of the clock and data inputs be maintained. As shown in Fig. 4.50, we require that the data input be stable around the clock event. If the data input changes while its value is being read, a bad value may be stored by the register. Most registers impose two constraints on the data stability:

- The **setup time** t_s is the time before the clock event for which the data input must be stable.
- The **hold time** t_h is the time after the clock event for which the data input must be stable.

We must choose the clock period such that all the events have propagated through the combinational logic to the outputs. Since all the registers receive the same clock, the clock period is determined by the worst-case delay for any of the combinational outputs.

The details of setup and hold time analysis are complex; the design of registers is usually left to experts. Both the access transistor and the internal amplifying gates play an important role. The speed with which the access transistor can be turned on and off depends in part on the characteristics of the clock waveform and in part on the parasitic elements around the switch.

We can use a small-signal model to help us understand the behavior of the gates in storing a value [Fla85; Sho88; Gin11]. In the circuit model of Fig. 4.51, the amplifying gates are modeled as an inverting ideal amplifier with gain A plus a resistance and capacitance: the resistance models the amplifier's internal resistance while the capacitance models the load presented by the other amplifier. If the internal

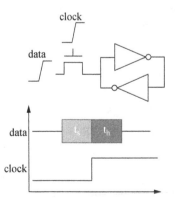

FIGURE 4.50

Setup and hold times.

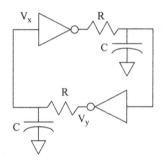

FIGURE 4.51

A simple analytical model for a register.

voltages are V_x and V_y, we can use the capacitor law to write formulas that relate the internal voltages:

$$C\frac{dV_x}{dt} = -\frac{1}{R}(AV_y - V_x),\qquad(4.17)$$

$$C\frac{dV_y}{dt} = -\frac{1}{R}(AV_x - V_y).\qquad(4.18)$$

We define $V = V_x - V_y$, which allows us to rewrite Eqs. (4.17) and (4.18) as

$$\frac{dV}{dt} = \frac{V}{RC}(A - 1).\qquad(4.19)$$

The solution to this equation is an exponential with time constant

$$\tau = \frac{RC}{A - 1}\qquad(4.20)$$

The time constant of the feedback amplifier system characterizes how quickly the register will amplify the incoming signal to a solid logic value—faster time constants mean that the register's hold time can be smaller. Since the time constant is inversely proportional to gain, higher gain in the inverters leads to faster transitions to the newly stored value. We will also use this analysis in our discussion of metastability in Section 4.4.4.

4.4.3 Clocking

Sequential machine timing

We need to determine the speed at which we can send clock inputs to the registers. Fig. 4.52 gives us a simple model for sequential machine clocking. Registers are connected to the signals at the boundary of combinational logic. Registers feed values into the combinational logic and hold the results of the combinational logic. The registers' storage holds the inputs steady and maintains the output value. A clock signal synchronizes the activity in the machine. During the course of one clock period, signals travel from registers at the combinational inputs, through the logic, and are stored

Sequential machine structure

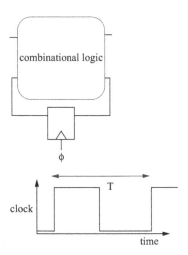

Clock signal

FIGURE 4.52

Sequential machines and clock period.

at the output registers. For simplicity, we will assume that the register in this example is a flip-flop with edge-triggered behavior.

The register's clock input must be activated (known as **clocking** the register) only when the combinational logic's output value has settled. If the combinational logic output changes after the clock event, it will not be stored in the register. The result will be an invalid machine state. Corrupt states, because they are recirculated by the machine's logic, cause permanent errors in the operation of the machine.

The machine's **clock period** T is a fundamental property of the machine. Clock period is the inverse of clock frequency:

$$f = \frac{1}{T}. \tag{4.21}$$

Clock period is a fundamental constraint on the correct operation of the machine. The clock period must be at least as long as the worst-case delay through the combinational logic; that longest delay is given by critical path analysis as we saw in Section 4.2.2. The critical path is the proper metric because it gives the worst-case delay from any input to any output. The sequential machine requires that all the combinational outputs be ready at once. All the logic primary outputs must have settled before we clock the registers.

If the worst-case delay for the combinational logic is Δ, then we require that

$$T > \Delta. \tag{4.22}$$

This is a **one-sided** timing constraint—it imposes a minimum bound on the clock period but no maximum bound. This means that we are guaranteed to find a clock

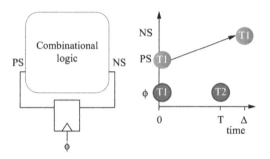

FIGURE 4.53

Errors caused by too-fast clocking.

speed at which the machine will operate—we can slow down the clock until the machine works properly. The one-sidedness of the machine's timing constraint is an important contributor to our ability to design reliable digital machines.

Fig. 4.53 illustrates what happens when the clock period does not meet the constraint of Eq. (4.22). The present state and primary input values are presented to the machine's combinational logic at time t. The logic's outputs are ready at time $t + \Delta$. However, the next clock event arrived earlier at time $t + T$. As a result, the registers load the wrong values.

The clock period allows us to map discrete time onto real time: sequential machine cycles $<0, 1, 2,...>$ occur at times $<0, T, 2T,...>$.

The clock period is also a fundamental metric of performance. A faster clock cycle means that the machine can perform more operations per second.

Setup/hold times and period

Realistic machines also incur some timing overhead in the registers due to setup and hold times. Taking these factors into account, the total clock period is

$$T \geq \Delta + t_s + t_h. \tag{4.23}$$

Pipelining

We can appreciate the importance of the discrete time model by studying **pipelining**. The original machine of Fig. 4.54 performs a complex function in one clock cycle—for example, multiplication. This machine does not have feedback to simplify our discussion, but pipelining can be extended to the feedback case. The clock period is determined by the delay through the logic: $T > \Delta$.

To pipeline the logic, we add a rank of registers in the logic. Let us assume that we can place the registers so that each logic block has the same delay $\Delta/2$. Properly inserting registers in the logic does not change the function it computes, only its clock cycle behavior. We want to insert registers to form a cutset through the combinational logic, similar to our approach to timing optimization. We refer to this set of registers as a **rank**. The clock period of the system is now constrained by $T > \Delta/2$—we have doubled the clock speed of the logic. Computer system performance can be measured

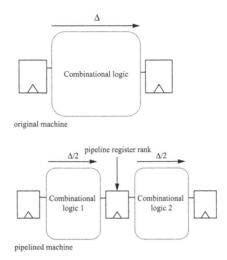

FIGURE 4.54

Pipelining.

using several metrics. In this case, we need to carefully distinguish between the effect of pipelining on two metrics:

- **Throughput** measures the number of results the machine produces per unit time. Since this machine produces one output per clock cycle, we have doubled its throughput by adding one rank of pipelining registers.
- **Latency** measures the time from input to output. Pipelining does not reduce latency; as we saw in Eq. (4.23), the registers will add a small amount of delay overhead.

Pipelining cannot improve latency because we cannot change the function performed by the combinational logic. But it can improve throughput because it allows the logic to be more efficiently utilized without breaking the assumptions of sequential machines. A sequential machine takes in one set of inputs each clock cycle to generate one set of outputs—each block of logic between two ranks of registers can operate on only one set of values at a time. When we add a rank of pipelining registers, we break the logic into two pieces, each of which can operate in parallel.

Register type and clocking

The type of register used affects the structure of the timing constraints and sequential machine. Consider the sequential machine of Fig. 4.55 built with a latch rather than a flip-flop. If the delay from the present state input to the next state output is shorter than the delay through other parts of the combinational logic, the next state value will be presented to the latch input before the end of the clock period. Because the latch is transparent, that value will flow through to the latch's output, where it will reenter the combinational logic. If the clock period is long enough, that new value could recirculate back to the latch's input, causing the wrong value to the stored.

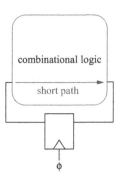

FIGURE 4.55

Problems with a single rank of latches.

For this machine to work, the clock period must satisfy a **two-sided timing constraint**:

$$2\Delta_{min} > T > \Delta_{max} \tag{4.24}$$

The clock period must be longer than the worst-case combinational delay but also shorter than the time required for changes to propagate back through the machine. Depending on the relationship between the shortest and longest paths through the logic, there may not be a feasible clock period that satisfies both constraints.

To avoid this two-sided constraint, we can build a **two-phase** machine as shown in Fig. 4.56. The combinational logic is divided into two pieces, each with a latch at its output. Each latch is controlled by a separate clock phase; the phases are designed such that at least one clock is always low. This nonoverlapping condition means that the latches will never both be open at the same time and there will never be a complete closed path through the logic. The length of each phase depends on the worst-case delay through the logic that feeds that phase's latch; the total clock period is the sum of these phase lengths plus the nonoverlapping intervals.

Clock skew   In order for the clock period T to be meaningful, we need to ensure that the clock signal itself arrives to all the registers at the same time. Since the clock signal acts as a global time base for the sequential machine, variations in the time at which the clock arrives at registers cause those registers to have a different notion of time, resulting in timing violations.

Fig. 4.57 shows a pair of registers. Both are connected to the clock φ, but reg1 receives the clock after a delay δ. Differences in the delay of the clock signal along different paths is known as **clock skew**. Assume that the clock signal φ is sent at time t_ϕ. The value is released from reg1 at release time $t_{rel} = t_\varphi + \delta$. Reg2 expects the combinational outputs at storage time $t_{stor} = t_\varphi + T$. The time available for the

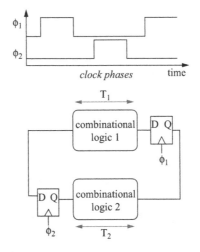

FIGURE 4.56

Two-phase, nonoverlapping clocks for latches.

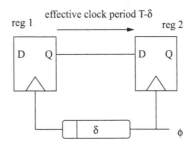

FIGURE 4.57

Clock delay and skew.

combinational logic to operate is $t_{stor} - t_{rel} = T - \delta$. The clock skew has reduced the effective clock period for this logic.

Fig. 4.58 shows a timing diagram for this situation. The skew of the clock to reg1 causes reg1 to release its data late. If the clock of reg2 was delayed by the same amount, the relative timing of these two registers would not change. But when reg2's clock is not delayed, the result is that the value transmitted from reg1 to reg2 has less time to be propagated through the machine.

It is important to remember that the effects of this error are not temporary. An improper value stored in the register will be used as input to the logic in the next cycle. That corrupted will continue to circulate throughout the system. Corrupted values in registers are **permanent faults**.

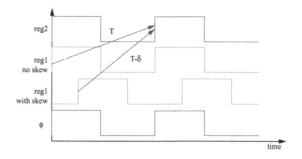

FIGURE 4.58

The effect of clock skew on system timing.

Example 4.6 Clock Skew

A combinational logic block has a maximum delay of 5 ps. We choose the sequential machine's clock cycle to be 5 ps, but we also design the circuit such that the input registers suffer a clock skew of 1 ps. Since the time from clocking of the input register to the clocking of the output register is only 4 ps transit time, values traveling through the combinational logic arrive late, causing errors.

4.4.4 Metastability

We must also consider what happens if the setup and hold times on a register are not met. Some basic physical principles show that not only will an intermediate value be stored in the register, but that the register may hold a **metastable** value for a very long time.

Metastability is an important source of errors in computer systems [Cha73]. Like most errors related to registers, it causes permanent faults in the operation of the machine. Metastability is particularly common at the boundary between machines that operate from different clocks. As chips become bigger, it is increasingly common to build complex systems out of machines with independent **clock domains**.

Fig. 4.59 shows the transfer curves for a pair of cross-coupled inverters in a register like that of Fig. 4.48. For simplicity, this plot centers the power supply voltages around zero. The horizontal axis is the value of x while the vertical axis is the value of y. This allows the same plot to show the transfer curves of both inverters: the top inverter uses the horizontal axis for its input and the vertical axis for its output; the bottom inverter uses the vertical axis for its input and the horizontal axis for its output. The curves intersect at three points. Two of those points are at the extremes of their ranges: $x = +v$, $y = -v$ represents x as logic 1, y as logic 0; $x = -v$, $y = +v$ represents x as logic 0, y as logic 1. The third intersection of the curves appears at the origin, where both x and y are at the midrange of the power supply. This is a metastable

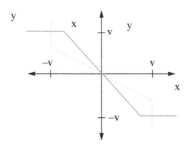

FIGURE 4.59

Metastability in cross-coupled inverters.

point—the system can attain and stay at that state for some time, but disturbances will cause it to move one of the other, stable operating points.

We can transform the cross-coupled inverter model of Fig. 4.51 to have the form of Fig. 4.59. The energy expended by each of the cross-coupled inverters is $^1\!/_2\, CV^2$ so the kinetic energy of the pair is

$$KE \;=\; \frac{1}{2}CV_1^2 + \frac{1}{2}CV_2^2 \qquad\qquad (4.25)$$

The system energy is the sum of the kinetic and potential energy. If we plot the register's potential energy as a function of the sum of the inverter voltages, it has the form shown in Fig. 4.60. The system states at both low and high voltages are stable because a small disturbance to the state causes it to return to a local minimum in energy. However, local minima must be separated by maxima. Those local maxima are not stable because a disturbance will push the system to one of the local minima states. But the system may stay at that local maximum for a very long time, much as a ball may stay at the top of a hill for a long time before rolling to the bottom.

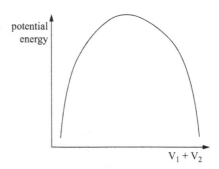

FIGURE 4.60

Energy versus system state for a register.

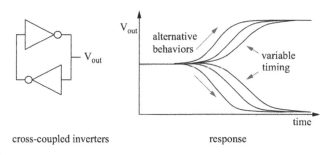

FIGURE 4.61

Metastable exponential divergence.

The behavior of a register when it reaches the metastable point is illustrated in Fig. 4.61. The output voltage can take either of two paths, one toward V_{DD} and the other toward V_{SS}. Which path it takes depends on the details of its internal state and environmental conditions. The circuit does not reliably take one path over the other. Its movement toward its ultimate destination grows exponentially—once it takes off in a direction, it moves quickly.

The fact that the register may ultimately stabilize at the wrong value is only part of the problem. The time that the register takes to stabilize to a valid logic value can vary over a huge time span. The temporal uncertainty of metastability means that we will always have some chance of the register not stabilizing no matter how long we wait.

Metastability is fundamental

It is important to keep in mind that metastability is an inherent result of the way in which we design computers—we can minimize it but not eliminate it. We use continuous physical quantities to represent discrete values; we use energy barriers to separate the states representing different discrete values. The energy barrier necessarily creates a metastable region between the stable regions.

Modeling

We can model the probability of a metastability failure of a register as

$$P_F = P_E P_S \qquad (4.26)$$

where P_E is the probability of storing a metastable state in the register and P_S is the probability that the register will not resolve the value to a valid state before it is used.

As shown in Fig. 4.62, a register is vulnerable to storing a metastable value during its setup and hold interval. We can model the probability of the register's input being metastable as being uniformly distributed over the clock period Δ, so the probability of storing a metastable value is

$$P_E = \frac{t_{SH}}{T} \qquad (4.27)$$

The probability of the register not resolving in the allowed time can be modeled as a Poisson process [Swa60]:

$$P_S = e^{-S/\tau} \qquad (4.28)$$

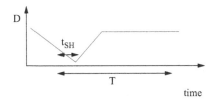

FIGURE 4.62

The vulnerable interval for metastability.

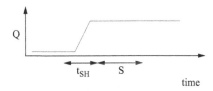

FIGURE 4.63

The stabilization window.

The time in which the register must resolve the metastable value is S. Fig. 4.63 shows that the stabilization window is after the setup/hold window. The stabilization window depends should be small relative to the clock period since the value must be resolved before it can be used; long stabilization windows reduce the chance of error but also cut into the available clock window since the value cannot be used until the end of the window.

Our analysis of feedback in registers helps us to understand the resolution failure process [Gin11]. We saw that the voltage difference between the outputs of the coupled inverters has an exponential form:

$$V = Ke^{-t/\tau}. \tag{4.29}$$

The time constant of Eq. (4.20) gives the speed with which the register will amplify a value from the metastable region to a stable value. This time constant is not always provided by the register circuit designers; we may need to estimate it to calculate failure rates. However, high-gain amplifiers in the register lead to short time constants that reduce the time required to resolve a metastable value.

When we combine Eqs. (4.27) and (4.28), the probability of a metastability failure is

$$P_F = \frac{t_{SH}}{T} e^{-S/\tau} \tag{4.30}$$

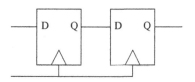

FIGURE 4.64

A dual-register synchronizer.

Example 4.7 Metastability Failures

Our clock rate is 1 GHz, and the setup/hold time of our registers is 10 ps. If we allow $S = 0.5$ ns to resolve a metastable value and our settling time $\tau = 10$ ps, then

$$P_F = \frac{0.01 \text{ ns}}{1 \text{ ns}} e^{-0.5 \text{ ns}/0.01 \text{ ns}} = 1.93 \times 10^{-24}$$

However, remember that this is the failure probability for a single register read. At a 1 GHz clock rate, the register performs 10^9 register operations per second. This still gives us a probability of failure over 1 s of operation as $P_{F,1G} = 1.93 \times 10^{-15}$.

We cannot eliminate metastability failures, but we can make them less likely to happen. We can reduce the probability of a synchronization failure is shown in Fig. 4.64. The value is read into one register, then into another before it is used. Resampling the data value significantly extends the value of S: first register provides a full clock cycle for resolution, much longer than in the single-register case. The sequential machine design must be adjusted to take into account the extra clock cycle of delay.

Taking action to minimize metastability is most important at the boundaries between regions with different clock periods. Blocks of logic may run at different clock rates for several different reasons: power management may reduce the clock speed in one part of the chip but not another; I/O systems typically run at slower clock rates to reduce their cost; and as we will see in Chapter 5, we may not be able to distribute the clock over the entire chip at a high enough speed, resulting in islands of logic running on their own clocks.

4.5 Synthesis

- The event model and gate-level delay models for combinational logic allow us to abstract the behavior of digital circuits.
- The gain of logic gates helps restore logic levels.
- Gain improves the dynamic response of logic signals.
- Power supply networks are subject to noise that can affect the operation of digital logic.

- Long wires are modeled as transmission lines for which voltage is a function of both space and time.
- Sources of noise include signal coupling, power supply, and input/output coupling.
- Synchronous machines use clocks to govern the times at which values are computed. Registers supply state memory.
- Metastability is a fundamental problem in the design of clock-controlled registers.

Questions

Q4-1   Give a table that shows the times at which events occur at the output of each gate in this logic network under each of these scenarios:

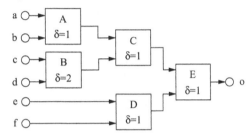

a. All inputs a–f receive events at $t = 0$.

A	1
B	2
C	2, 3
D	1
E	2, 3, 4

b. Inputs b and c receive events at $t = 0$, input e receives events at $t = 1, 2, 3$.

A	1
B	2
C	2, 3
D	1, 2, 3
E	2, 3, 4

c. Inputs a and c receive events at $t = 0, 2$, input e receives events at $t = 2, 3$.

A	1, 3
B	2, 4
C	2, 3, 4, 5
D	3, 4
E	3, 4, 5, 6

Q4-2   For this logic network:

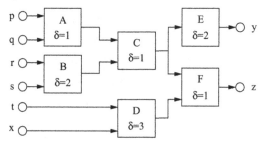

a. Find the forward weights for delay analysis.
b. Find the backward weights for delay analysis.
c. Find the critical path.

Q4-3   For this network:

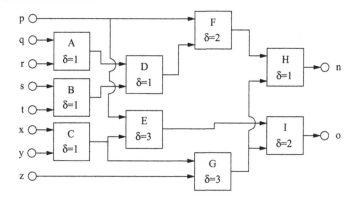

d. Find the forward weights for delay analysis.
e. Find the backward weights for delay analysis.
f. Find the critical path.

Q4-4   A set of gates has $V_{IL} = 0.25$ V, $V_{IH} = 0.6$ V, $V_{OL} = 0.1$ V, $V_{OH} = 0.85$ V. How much ground bounce would be required on a gate to make it unable to send a logic 0 to another gate?

Q4-5   A block of logic contains 1000 gates each with $R_t = 9$ kΩ, $C_L = 1.8$ fF. The power supply voltage is 1 V.
    a. What is the maximum current that each gate can draw based on the resistive model?
    b. If all gates draw their maximum current for 20 ps, what is the decoupling capacitance required to protect against a 10% power supply swing?

Q4-6   A block of logic operates at a power supply voltage of 1 V and can tolerate a power supply bounce of 10%. Each gate has $R_t = 20$ kΩ, $C_L = 3$ fF.
    a. Plot decoupling capacitance for $t = 1$ ns, $100 \leq n \leq 10{,}000$.
    b. Plot decoupling capacitance for $n = 1000$, 0.1 ns $\leq t \leq 10$ ns.

Q4-7   A chip draws 10 A of current over a 1.1 V power supply. Pin inductance is 2 nH. How many V_{DD} pins are required to limit the maximum voltage drop across the pins to 0.1 V when current changes from 0 to the maximum value in 1 ns?

Q4-8   Prove that the resistance of a wire on a chip can be measured in units of ohms/square using the relationship between R (resistance) and ρ (resistivity).

Q4-9   The resistivity of copper at room temperature is 17E-9 Ωm. The on-chip wire has a thickness of 20 nm.
    a. What is its sheet resistance?
    b. What is the resistance of a wire of length 1 μm, width 20 nm?

Q4-10   Consider a set of wires with coupling capacitance and capacitance to ground. Both wires have the same uniform width W $= 20$ nm and height H $= 15$ nm. Unit ground capacitance is 30 aF/μm^2 and unit coupling capacitance is 0.35 aF/μm. Sheet resistance is 5 ohms/square. Plot ground and coupling capacitance (separate curves) for the range of lengths 1 μm $<$ L $<$ 100 μm.

Q4-11   Consider several wires with coupling capacitance and capacitance to ground. Unit ground capacitance is 20 aF/μm^2 and unit coupling capacitance is 0.9 aF/μm. The width is 10 nm wide. The process sheet resistance is 5 ohms/square. Calculate Elmore delay for each wire; use five sections for your Elmore model.
    a. L $= 1$ μm.
    b. L $= 10$ μm.
    c. L $= 50$ μm.
    d. L $= 1$ mm.

Q4-12   A wire has unit capacitance is 10 aF/μm^2 and a sheet resistance of 5 Ohms/square. The wire is 10 μm long and 20 nm wide. Plot Elmore delay for $2 \leq n \leq 10$.

Q4-13 A fabrication process gives a unit capacitance is $10 \text{ aF}/\mu\text{m}^2$ and a sheet resistance of 5 ohms/square. A wire fabricated in this process has three series connections: section 1 has $W = 50$ nm, $L = 100$ nm; section 2 has $W = 25$ nm, $L = 300$ nm; section 3 has $W = 15$ nm, $L = 600$ nm. What is its Elmore delay? Assume one section per wire segment.

Q4-14 In Fig. 4.57, what is the effective clock period for reg2 if it receives the clock signal ϕ after reg1?

Q4-15 You are given transistors with $R_n = 12 \text{ k}\Omega$, $R_p = 25 \text{ k}\Omega$, $C_L = 2 \text{ fF}$. The clock period T for your system is equal to the delay of a chain of n inverters. If effective resistance and load capacitance vary by $\pm 20\%$, what are the best-case and worst-case clock periods for the system as a function of n? Use an RC model and 50% delay to estimate delay.

Q4-16 Plot the value of N_p for Rent's Rule for $100 \leq N_g \leq 10,000$ for two sets of parameters:
  a. Rent's classic parameters $r = 0.6$, $K_p = 2.5$
  b. Microprocessor parameters $r = 0.45$, $K_p = 0.82$

Q4-17 A microprocessor chip in 1974 had 1500 gates and 40 pins.0
  a. If $r = 0.6$, $K_p = 2.5$, how many pins does Rent's Rule predict that the chip should need?
  b. If $r = 0.45$, $K_p = 0.82$, how many pins does Rent's Rule predict that the chip should need?

Q4-18 A clock has a period of 1 ns. Combinational logic's primary inputs are connected to register A, and the logic's primary outputs are connected to register B. The clock arrives 0.1 ns late at register A; the clock suffers no delay to register B. What is the maximum allowable combinational logic delay, ignoring setup/hold times?

Q4-19 A register has a combined setup/hold time of 0.005 ns. The system clock period is 0.5 ns. Assume $\tau = 1$ ps and that the resolution time S is 0.1 ns. What is its metastability probability of failure?

Q4-20 A register has $t_{setup} = 0.01$ ns, $t_{hold} = 0.01$ ns. The system clock period is 0.5 ns. Assume $\tau = 10$ ps. Plot the $\log P_f$ for $100 \text{ ps} \leq S \leq 1$ ns.

Processors and Systems

5

5.1 Introduction

Now that we understand how to build sequential machines, we are ready to design complete computer systems: processor, memory, and the interconnections between them. We will concentrate not on the instruction set but on the physical aspects of computer systems and how they affect performance, energy consumption, and reliability. We will see that some physics-based problems in computer design have physics-based solutions while others are solved using other means.

Fig. 5.1 shows the organization of a typical computer system:

- The CPU or processor performs computations.
- The memory stores both instructions and data.
- I/O devices provide input and output.
- The bus interconnects all the components.

This block diagram draws on both the Turing and von Neumann models: the Turing machine provides the processor and memory playing the role of the tape; the von Neumann model describes the CPU and memory, with the bus transmitting the address, data, and control signals required for their operation. I/O devices allow us to talk to the machine and view its operation.

Computer systems are built for a wide range of applications. In this chapter, we will study two ends of the spectrum: mobile and server systems.

Mobile systems (smartphones, tablets, laptops, etc.) often deliver surprisingly high performance but must consume as little energy as possible. Their batteries provide a finite amount of energy; higher energy per operation means shorter battery life. Power consumption is also important in these systems since they are constrained in their thermal capabilities. Many modern laptops have fans to help cool the electronics but smartphones do not.

Server systems operate at fixed locations and draw their power directly from the electric grid. A server may run at very high power levels. As a result, its thermal characteristics are of critical importance.

We are now in a position to see how the properties of our components affect the behavior of complete computer systems. As always, we are interested in performance,

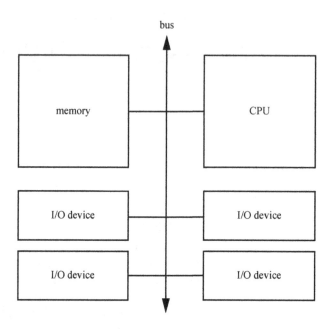

FIGURE 5.1

Block diagram of a computer system.

power, and reliability. We will start with a brief introduction to some concepts in system reliability. Section 5.3 looks at CPUs while Section 5.4 studies memories. Section 5.5 explains the operation of mass storage devices and their implications for performance. Section 5.6 evaluates system power consumption for both server and mobile systems. Section 5.7 studies heat transfer and the relationship between heat and reliability.

5.2 System reliability

Even though the error rate for CMOS gates is very low, systems built from many components require extremely high probabilities to achieve reasonable levels of system reliability. Sieworek and Swarz [Sie82] describe basic principles of computer system reliability modeling.

We generally model the basic reliability of a component as the **failure rate** $f(t)$ or probability of failure per unit time. From this basic failure rate, we can compute the **cumulative failure rate** over a longer interval from time 0 to t:

$$F(t) = \int_0^t f(t)dt \qquad (5.1)$$

The reliability function is

$$R(t) = 1 - F(t). \tag{5.2}$$

$R(t)$ gives the probability that no failure will occur in the interval $[0,t]$. We also characterize systems by their **hazard function**

$$z(t) = \frac{f(t)}{1 - F(t)} \tag{5.3}$$

$z(t)$ gives the instantaneous number of failures per unit time.
The mean time to failure is

$$MTTF = \frac{total\ time}{\#\ failures} \tag{5.4}$$

$MTTF$ is often measured experimentally.

A simple probability model for digital systems can be built by assuming the unit of time to be a single clock period. A failure of a component at any time during that clock period causes a system failure. In this model, a device either fails or is valid:

$$P_{err,dev} + P_{OK,dev} = 1 \tag{5.5}$$

where $P_{err,dev}$ is the probability of failure of a device in one clock cycle and $P_{OK,dev}$ is the probability that the device does not fail in that clock cycle. The system fails on a clock cycle if any device fails. The probability of a single device being valid in a clock cycle is

$$1 - P_{err,dev} \tag{5.6}$$

The probability of all n devices being valid during the clock cycle is

$$P_{OK,sys} = \left(1 - P_{err,dev}\right)^{n} \tag{5.7}$$

This means that the probability of an error for the n-device system on one clock cycle is

$$P_{err,sys} = 1 - \left(1 - P_{err,dev}\right)^{n} \tag{5.8}$$

Since Eq. (5.8) only describes the failure rate for one clock cycle, we should also consider system reliability over many clock cycles. By making use of clock cycle–based failure probabilities, we can reduce the reliability calculation to a discrete form in which the system does not fail for every cycle in the interval $[0,N]$. The reliability of the system over N clock cycles is

$$R_{sys}(N) = \left(1 - P_{err,sys}\right)^{N} \tag{5.9}$$

The form of the reliability equation suggests that the mathematics of computer system reliability is daunting—the huge number of operations performed by a computer means that each component must be extremely reliable for the system to avoid failing at an unacceptable rate.

Example 5.1 Device and System Error Rates

One of the first microprocessors, the Intel 8008, had about 3400 transistors. Let us assume that the probability of error for a single transistor in one of these microprocessors was 1×10^{-6}. This error rate is much higher than the actual error rate but serves to prove a point. Under this assumption, the error probability of the chip for one clock cycle would be

$$P_{err,sys} = 1 - \left(1 - 10^{-6}\right)^{3500} = 0.0035$$

The system error rate is substantially higher than that of the components. We often refer to reliability using **nines**. In this case, the device has 5–9 s reliability while the chip, with a reliability level of 0.9965, has only 2–9 s reliability. If the transistor error rate is 1×10^{-9} (9–9 s), the probability of an error increases to

$$P_{err,sys} = 1 - \left(1 - 10^{-9}\right)^{3500} = 3.49 \times 10^{-6}$$

The 8080 ran at a clock rate of 800 kHz. At that clock rate, the reliability over 1 s is 0.94; equivalently, the probability of a failure in that interval is 0.06.

Given that modern chips have billions of devices and run at clock rates of billions of times per second, we need extremely reliable devices to create systems with reasonable levels of reliability. Now consider a modern chip with a billion transistors, each with a probability of error of 1×10^{-9}. This gives an error probability for one cycle of 0.6321. If the chip runs at 1 GHz, its reliability value for 1 s would be approximately 0. Substantially higher device reliability values are required to give reasonable system reliability levels.

Highlight 5.1

Reliable systems require extremely reliable components.

Bathtub curve

Failure rates of electronic components follow a characteristic shape known as the **bathtub curve**. As shown in Fig. 5.2, failure rates are initially high, then stabilize at a low level for a long period, and then rise again. Early-lifetime failures are known as **infant mortality**. Natural manufacturing variations result in some components with marginal properties that will cause them to fail within a few hours of operation. Burn-in procedures allow manufacturers to catch these cases before they are shipped to customers. After a period of use, a variety of physical **aging** effects cause failure rates to rise. The length of the low failure rate interval depends on the design of the components—consumer-grade devices are typically designed to shorter lifetime expectations than are industrial-grade components. We will discuss the role of heat in reliability and aging in Section 5.7.3.

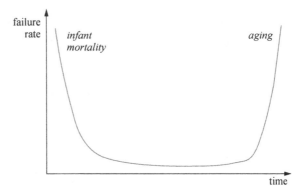

FIGURE 5.2

The bathtub curve for failure rates.

5.3 Processors

The CPU is a complex machine. But some basic physical principles are critical to an understanding of processor design. After reviewing some characteristics of modern processors, we will study three critical problems in CPU design: busses, global communication, and clock distribution networks.

5.3.1 Microprocessor characteristics

The ITRS Roadmap demonstrates some several key trends at the chip level. As shown in Table 5.1, the roadmap is divided into low-cost and high-performance segments. We can see that power consumption does not scale as much as predicted according to ideal scaling. Clock frequencies scale less aggressively for low-cost chips than for high-performance. An important difference between low-cost and high-performance chips is the number of pins that must be provided. We will see that supplying power on-chip is one of the important drivers of high pin counts. As we saw in Section 4.3.4, Rent's Rule parameters become less aggressive once system integration with sufficiently high levels. The values $r = 0.45$, $K_p = 0.82$ model the microprocessor case. Memory interfaces and busses require fewer pins than do the random logic for which Rent's Rule was originally formulated. High levels of integration have eased the requirements on package pinout relative to what they would be for random logic at billion-transistor levels of integration.

Modern microprocessors exhibit complex memory hierarchies. Even inexpensive embedded processors often include a cache and onboard flash storage. The memory hierarchies for high-performance microprocessors are even more complex. Fig. 5.3 illustrates a complex memory hierarchy with relative sizes and access times (referenced to a 1 GHz clock). Even within the span of SRAM and DRAM, we see a performance range of 80:1 and a size range of 3×10^7:1. When we take into account

Table 5.1 Packaging Trends From the ITRS Roadmap [ITR13]

	2011	2014	2017	2020	2023
Cost-performance chips					
Power (W)	161	152	130	130	130
I/O count	728–3061	800–4075	960–5423	1050–7218	1212–8754
Performance (GHz)	3.744	4.211	4.737	5.329	5.994
High-performance chips					
Power (W)	161	152	130	130	N/A
I/O count	5094	5896	6826	7902	9148
Performance (GHz)	3.744	4.211	4.737	5.329	5.994

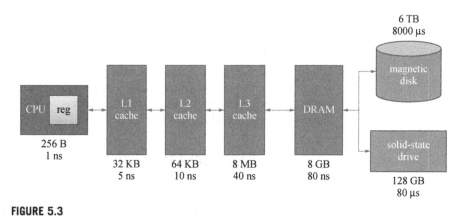

FIGURE 5.3

Relative access times across the memory hierarchy.

mass storage devices, which are used for virtual memory as well as file storage, the range of sizes and access times become even larger. Memory hierarchies are complex because we cannot build a single memory structure that is ideal across all axes: size, performance, and volatility. Complex memory hierarchies are the result of adaptations made by computer architects to the limitations of semiconductor technology.

A **floorplan** describes the physical design of a chip at a high level of abstraction. As shown in Fig. 5.4, the units in the floorplan design are equivalent to the system block diagram components, although the division of the design into components may vary somewhat between the logical and physical designs. A floorplan should be drawn roughly to scale. The relative sizes of the blocks are important, as are their width/length aspect ratios. A floorplan helps chip designer to budget their design, particularly when the maximum chip area must meet a given specification. The die photo shown in Fig. 5.5 shows that the varying structures of different modules on the chip are visible at a large scale.

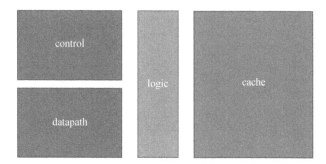

FIGURE 5.4

A chip floorplan.

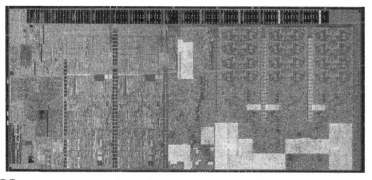

FIGURE 5.5

Die photo of a Broadwell processor.

Courtesy Intel Corporation.

We next consider two subsystems of the CPU that have a major influence over both performance and power system: busses and clock distribution networks. Both of these subsystems are influenced by the organization of the chip floorplan.

5.3.2 Busses and interconnect

The bus in a computer system connects all the major components, allowing them to communicate with each other. A small bus is used in the datapath to connect all the function units and register file. Much larger busses connect the processor to the memory and I/O systems. Many modern chips use networks-on-chips that provide more complex interconnection topologies and use packet-based communication. The physical principles of design are similar in both cases, and we will concentrate on busses for simplicity.

We can gain an intuition for buffering with a simple model of a long wire illustrated in Fig. 5.6. A wire segment has a section of wire terminated at one end by a

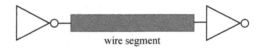

wire segment

circuit model

FIGURE 5.6

A model for buffers on long wires.

driver inverter and the other end by a receiver. The driver is modeled as a driving resistance R_b and the receiver by a load capacitance C_b. The wire itself is described by the Elmore RC model. When we cascade a series of buffered wire section together, the receiver for one segment becomes the driver for the next.

Although the model for the middle sections of the wire are standard RC sections, we have to take into account the driver and receiver at the ends. The first section has the form

$$(R_b + r)c \tag{5.10}$$

where the wire section impedances are r and c. The last section of the wire containing the load capacitance has the form

$$(nr + R_b)(C_b + c) \tag{5.11}$$

The middle segments of the wire $[2, \cdots, n-1]$ are RC sections. The complete buffered wire segment delay can be approximated by

$$
\begin{aligned}
t_{seg} &= (R_b + r)c + \sum_{2 \le i \le (n-1)} (n-i)rc + (nr + R_b)(C_b + c) \\
&= (R_b + r)c + \frac{1}{2}(n-1)(n-2)rc + (nr + R_b)(C_b + c) \tag{5.12}
\end{aligned}
$$

For short wires, the buffer resistance and capacitance will dominate the delay. As the wire grows longer, the wire delay becomes a significant factor.

We want to compare the delay of one long segment versus two short segments as shown in Fig. 5.7. A fair comparison requires that each case has the same total transistor area A. When we put a buffer in the middle of the wire, each driver has a transistor active area of $A/2$. Each half-length segment has $n/2$ RC sections and approximately 1/4 the delay of the full length segment.

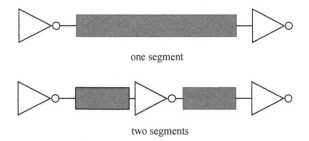

one segment

two segments

FIGURE 5.7

A long wire divided into one versus two segments.

Since the two half-length segments are in series, the total delay of the buffered wire is half that of the unbuffered version. In this case, the total delay of the buffered wire is the sum of the delays of its two buffered segments. In contrast, when we directly connect two RC transmission lines, we cannot add the Elmore delays of the two segments to find the total wire delay. Elmore analysis tells us that the capacitance of a section of the second segment is charged through the resistance of both sections of wire. When we use an inverter to provide a buffered connection between the wires, the buffer serves as an endpoint for the chain of resistances through which a segment's capacitance is charged. That buffer amplifies and reshapes the transitions along the wire, increasing their slope and reducing their distortion. We can therefore treat the delays of the two buffered segments as independent and add their delays.

We can modify Eq. (5.12) for the case of a buffered wire. If N is the number of buffered segments, then each segment has n/N RC sections, each of which is the same size as one of the RC sections in the original wire. The delay of one segment is

$$t_{bseg} = (R_b + r)c + \frac{1}{2}\left(\frac{n}{N} - 1\right)\left(\frac{n}{N} - 2\right)rc + \left(\frac{n}{N}r + R_b\right)(C_b + c) \qquad (5.13)$$

And the total delay of the buffered wire is

$$t_{bwire} = N\left[(R_b + r)c + \frac{1}{2}\left(\frac{n}{N} - 1\right)\left(\frac{n}{N} - 2\right)rc + \left(\frac{n}{N}r + R_b\right)(C_b + c)\right] \qquad (5.14)$$

For moderate numbers of buffers, the delay depends largely on the Elmore delay of the wire segments; since each of the buffered segments is much shorter than the original wire, the buffers improve delay. As we add more buffers along the wire, the impedance and delay of the buffers come to dominate so that total delay eventually increases. This means that we can identify an optimal number of buffers to minimize wire delay. Fig. 5.8 shows total delay as a function of the number of buffers along the wire. This simple analysis ignores several effects: the signal coming into the buffer will have a slow rise time, increasing the buffer's delay; realistic wires will have inductive effects; etc. But this model suggests that buffering to restore the signal on long wires provides significant benefits.

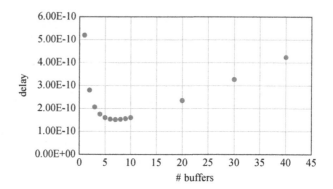

FIGURE 5.8

Delay versus number of buffers.

The logic attached to the bus also has a significant effect on its performance. We can estimate the delay through a bus as a function of the complexity of the system that it connects [Ser07]. As shown in Fig. 5.9, a bus connects N units known as **cores**. If we normalize lengths to the size of a core, assuming that all cores are the same size, then the length of the bus is proportional to the number of cores it connects. The bus is used by one core to send values to another core. Each core has to be able to drive the bus at the required speed, which requires large driver transistors. The capacitance of the bus is one factor that determines the required size of the drivers. However, the capacitive load of the cores themselves is another important factor. Each core has a receiver circuit, shown here as a register. The receiver circuit has capacitance on its input. That capacitance is always attached to the bus, even if the core is not actively listening to the bus. Each of the cores has to be able to drive both the bus wire capacitance and the loads presented by the cores.

The delay along the bus has two major components: the delay of the drivers used by the core to send the value on the bus and the delay along the bus line itself:

$$\delta_{bus} = \delta_{in} + \delta_{wire} \qquad (5.15)$$

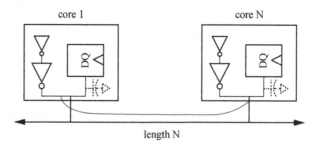

FIGURE 5.9

Model for delay on a bus.

We model the driver delay using the formula for driving large loads (using k for the number of driver stages):

$$\delta_{in} = k\left(N\frac{C_L}{C_g}\right)^{1/k} \qquad (5.16)$$

We model the delay along the bus line using Elmore delay (assuming that the delay has no repeaters):

$$\delta_{wire} = \frac{1}{2}rcN(N-1) \qquad (5.17)$$

We can rewrite the formula for bus delay as

$$\delta_{bus} = k_1(C_LN)^{1/k} + k_2N^2 \qquad (5.18)$$

where k_1 and k_2 are constants derived from the component formulas. This formula shows that bus delay depends on the square of the number of cores that the bus connects.

Chip-to-chip communication can be a bottleneck in many systems. PCI Express (PCIe) [Bud04; Int14] is widely used for system-level interconnection. PCIe interconnection networks are not based on busses; instead, it uses point-to-point links with switches to interconnect multiple devices. The links can be built in a variety of widths: $\times 1$, $\times 2$, $\times 4$, $\times 8$, $\times 12$, $\times 16$, or $\times 32$. A link is full duplex with simultaneous bidirectional connection between the two devices connected by the link. Communication on the link is divided into packets. The physical layer of the link is optimized for very high-speed communication without distributing a clock between the devices. High-speed communication without an explicit clock requires that a clock be recoverable from the data, which requires that the physical layer bit stream transition at some minimum rate. PCIe uses an 8-to-10 bit code to ensure that a transition occurs at least every fifth bit.

5.3.3 Global communication

The speed at which data travel places a fundamental limit on the operation of computers. This is particularly true given the high clock rates at which modern microprocessors operate. The next two example looks at the velocity of data from different perspectives.

Example 5.2 Velocity of Data

Even if data travel at the speed of light in a vacuum, it does not travel far in the duration of a gigahertz clock cycle. Assume a clock period of

$$T = \frac{1}{f} = 10^{-9}\text{ s}$$

The distance that a pulse representing a bit can travel in one clock cycle, assuming speed-of-light transmission, is

$$d = cT = 3 \times 10^8\text{ m/s} \cdot 10^{-9}\text{ s} = 3 \times 10^{-1}\text{ m}$$

The velocity of data over an *RC* transmission line is much lower. Assume a wire width $= 0.1$ µm and wire length $= 10^4$ µm. Then:

$$r = 0.04 \ \Omega/\blacksquare \cdot (1 \ \mu\text{m}/0.1 \ \mu\text{m}) = 0.4 \ \Omega$$

$$c = 0.5 \ \text{aF}/\blacksquare \ \mu\text{m} \cdot (1 \ \mu\text{m}/0.1 \ \mu\text{m}) = 5 \ \text{aF}$$

$$\delta = \frac{1}{2} rcn(n+1) = \frac{1}{2}(0.4 \ \Omega)(5 \ \text{aF})10^4 \left(10^4 - 1\right) = 10^{-7} \ \text{s}$$

$$v = \frac{l}{\delta} = \frac{10^{-2}}{10^{-7}} = 10^5 \ \text{m/s}$$

A pulse on an *RC* transmission line can travel only a much shorter distance in one clock cycle:

$$d = vT = 10^5 \ \text{m/s} \cdot 10^{-9} \ \text{s} = 10^{-4} \ \text{m} = 100 \ \mu\text{m}$$

This simple model is somewhat pessimistic, but the propagation speeds of signals on wires means that data cannot traverse large chips in a single cycle at today's high clock speeds.

We can illustrate the challenges of distributing a clock signal over a large chip by considering the design of a planet-scale computer. Computer systems built to planetary scale have been imagined in science fiction, notably in the film *Forbidden Planet*.

Example 5.3 Planet-Scale Computer

NASA

NASA

We will use earth as a model for our planet-scale computer. Here are the earth's basic measurements:

Mean radius	6371 km
Volume	1×10^{12} km^3
Total area	510×10^6 km^2
Land area	149×10^6 km^2

For simplicity, we will assume that we fill our planet-scale computer with computers only to a depth of 10 km. This still gives a total machine room volume of 5×10^9 km^3. If we assume that each computer has dimensions 0.5 m $\times$ 0.1 m $\times$ 0.5 m = 0.025 m^3 = 25×10^{-9} km^3. This means that the planet-scale machine room holds 2×10^{20} computers. If each computer consumes 100 W, then the planet-scale computer consumes 2×10^{22} W. If we assume a clock rate of 1 GHz and one instruction per clock cycle, then the planet-sized computer can execute 200×10^{24} instructions per second. If we assume 1 GB RAM and 1 TB of disk, then the system includes 200×10^{24} of RAM and 200×10^{27} bytes of disk storage. That amount of disk storage can hold about 8×10^{15} high-definition movies.

While the capacity of a planet-scale computer is impressive, its speed is not. The system's data latency is ultimately limited by the speed of light. Even if we route data directly through the core of the planet-scale computer at the speed of light, the delay between opposite points on the surface is 42 ms. For comparison, the clock speed of the Intel 4004, the first microprocessor introduced in 1971, was 9 μs.

However, we can fix the clock distribution problem by redesigning our planet-sized computer to perform different types of jobs. The fully synchronous organization is well suited to a large, single computation. **Amdahl's law** recognizes that most such programs have parts that can be parallelized and parts that cannot, as illustrated in Fig. 5.10. It calculates the possible speedup based on the fraction of the job P that can be parallelized and the number of tasks N into which the job can be parallelized:

$$S(n) = \frac{1}{(1 - P) + \frac{P}{N}} \tag{5.19}$$

Much of today's computation is performed as a large number of small jobs, such as Web page access and database transactions. These tasks do not need to operate on the same clock, and each processor can run at a clock rate determined by its own internal physics, not the planetary-scale physics. The speed of light would still limit any communication that was required between tasks, as shown in Fig. 5.11.

5.3.4 Clocking

One of the key problems in large chips is the proper distribution of the clock signal. We saw in Chapter 4 that clock skew can create timing violations. We also need to be

FIGURE 5.10

Series and parallel components of a job for Amdahl's law.

sure that the transitions on the clock signal have sharp edges to be sure that transistors controlled by the clock turn on and off quickly. As illustrated in Fig. 5.12, a clock signal requires aggressive specifications for its rise time. Both rise/fall time and skew cut into the amount of time available in a clock period to perform computation.

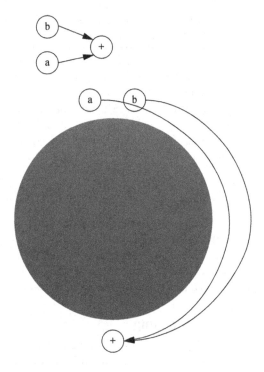

FIGURE 5.11

Communication delay between jobs.

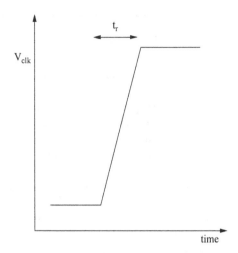

FIGURE 5.12

Rise time in a clock signal.

Clock signals are typically designed for a rise time of 10% or less of their period. A 2 GHz clock signal with a 10% rise time budget would give a rise time of 50 ps [Ish16].

Clock distribution faces two major problems. The first is simply the long distances that must be covered. A chip 1 cm on a side, for example, requires the clock be distributed from a single point over wires that are perhaps a million times longer than the length of the transistor channels.

The second problem may be less obvious—the clock signal includes a huge capacitive load. The clock signal is attached to thousands or tens of thousands of registers. Each register presents a capacitive load to the clock. In addition, the wires carrying the clock signal have their own capacitance. The huge capacitance makes it difficult to create the sharp clock edges required as well as requiring huge currents. On many chips, the clock is the single largest consumer of energy.

Example 5.4 Capacitive Loads on the Clock Network

The simplest form of the dynamic register has one transistor gate connected to its clock input. If $C_g = 0.9$ fF, then 10,000 registers would create a load of $9 \times 10^{-10} = 90$ nF.

If we assume the clock signal swings through 1 V, then the charge required on the clock network is $q_c = C\Delta V = 9 \times 10^{-10}$ F $\cdot 1$ V $= 9 \times 10^{-10}$ C. If the required rise time is 100 ps, then the current required to charge the clock capacitance is

$$I_c = \frac{q_c}{t_r} = \frac{9 \times 10^{-10} \text{ C}}{100 \times 10^{-12} \text{ s}} = 9 \text{ A}$$

Highlight 5.2

Clock distribution accounts for about 30% of total power consumption in a modern high-performance microprocessor.

The large capacitive load gives us a clue as to how to solve the problem—the exponentially tapered buffers used to drive large loads from Section 3.4.3. The difference between the two cases is that our original problem assumed a single large capacitance, such as the pin on a package, while in this case our capacitive load is distributed across the chip.

Because we need to send the clock to many different locations, we can organize the clock signal into a **tree**. (Some clock distribution networks use mesh interconnections, but these networks are harder to analyze.) As shown in Fig. 5.13, tree gives shorter connections from the clock's source to each register than would a single wire snaking around the chip.

Because the clock must be distributed across the two-dimensional chips, we must find a way to organize the clock tree to distribute the clock signal to all parts of the chip. One simple and effective organization is the H-tree, shown in Fig. 5.14, so named because it takes the form of smaller and smaller Hs. H-trees work best when all sections of the chip have about the same number of registers; it is also possible to create a specialized, unbalanced clock tree for chips with less even distribution of registers. Some chips also use mesh networks to distribute the clock.

Taking a cue from our optimal buffer sizing, we will add buffers to the clock tree. As shown in Fig. 5.15, a natural place to put each buffer is at the branch point of the tree. The **branching factor** b of the tree—the number of branches leaving a given branch point—determines the amount of capacitive load from buffers that each buffer

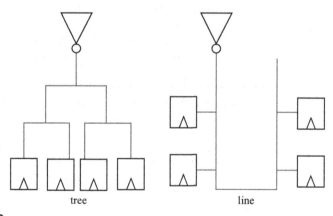

tree   line

FIGURE 5.13

Trees versus lines for clock distribution.

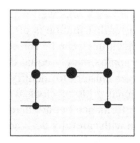

FIGURE 5.14

An H-tree for clock distribution.

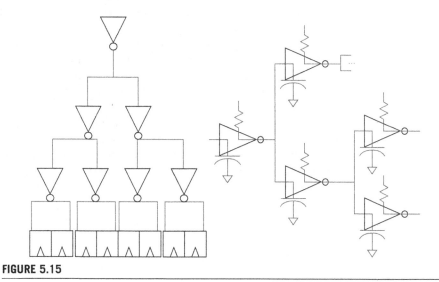

FIGURE 5.15

A buffered clock tree.

must drive. (We will, for simplicity, ignore the capacitance of the clock tree wires.) In this case, $b = 2$, so each buffer drives a load of $2C_L$.

If all the buffers in the tree have equal-sized transistors, the tree still creates an exponentially tapered buffer tree because each level has b times more buffers than the previous stage. The total number of buffers in each stage is b^i. In our exponentially tapered buffer problem, we assumed that each stage was α times larger than the previous stage. In the case of the clock tree, $\alpha = b$. Optimal driving still requires $\alpha = e$, but we cannot put an irrational number of buffers at a clock tree node. However, we can come close to the optimal case, with $b = 2$ or $b = 3$.

The buffers become larger as we move from the leaves to the root of the clock tree. If we assume that each register has an input capacitance equal to a transistor of minimum size, we can make the driver for the leaves b times larger. The drivers at each successive stage are larger by the same amount.

The H tree works well when the registers are evenly distributed across the chip. But register distribution is often uneven in large, complex chips. In this case, unbalanced clock trees can be formed by clustering together registers to form groups that present approximately equal capacitive loads. These clusters can then be combined into larger clusters to identify the structure of the unbalanced tree. Meshes—two-dimensional grids of clock wiring—are also used on some chips. Meshes provide good performance but are harder to analyze than trees.

Clock domains

Even with carefully designed clock distribution networks, getting the clock everywhere on chip is very challenging. Increasingly, it is physically impossible to distribute a global clock, as we saw in Example 5.2. As a result, large chips are not designed as a single synchronous machine. Instead, a design style known as **globally asynchronous locally synchronous (GALS)** allows parts of the chip to operate synchronously with carefully designed asynchronous communication at the boundaries between these synchronous units. As shown in Fig. 5.16, the chip is divided into clock domains, each of which is synchronous. When passing data between clock domains whose clocks are not synchronized, we use **synchronizers**. A synchronizer is a register used to communicate asynchronously. Metastability is a major problem at these boundaries.

The clock signal is generated off-chip and sent on-chip through a pin. Clock generators typically rely on crystal oscillators to deliver accurate and stable clock frequencies. A crystal oscillator relies on the **piezoelectric effect** which relates mechanical stress and charge. The equivalent circuit for a piezoelectric crystal is shown in Fig. 5.17 [Pec97]. C_1 and L_1 are the main impedances associated with the

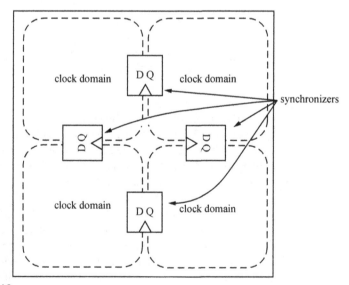

FIGURE 5.16

Clock domains on a large chip.

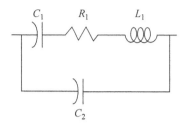

FIGURE 5.17

Equivalent circuit of a piezoelectric crystal [Pec97].

piezoelectric charges; R_1 represents losses and C_2 represents parasitic capacitances of the connections to the crystal. The series resonant frequency of the crystal is

$$f_s = \frac{1}{2\pi\sqrt{L_1 C_1}}. \tag{5.20}$$

Fig. 5.18 shows a crystal oscillator circuit known as a **Pierce oscillator**. The crystal's inductance and capacitance create a resonant feedback loop around the inverter.

Sending the clock signal into the chip also presents physical challenges. The clock signal must be delivered at high speeds and with low distortion. Pins present both inductance and capacitance which cause increased distortion at higher speeds. The off-chip clock can be generated at a lower frequency, such as 100 MHz, and injected into the chip. The chip includes a **phase-locked loop** (**PLL**) to generate the required internal clock frequency from the external clock. A PLL is an oscillator designed to track the phase of the input signal. The PLL allows the on-chip clock to mimic the stability of the off-chip clock but at a higher frequency.

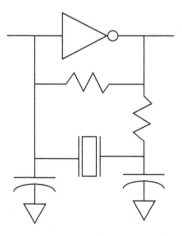

FIGURE 5.18

A Pierce oscillator.

5.4 **Memory**

Memory is a key component in the computer system; we saw in Chapter 1 that the early decades of computer development were consumed with the search for a physical device that could provide useful memory. Examining its operation and properties is essential to the understanding of computer systems.

Memory continues to grow in size, but fundamental speeds have not significantly increased. Table 5.2 shows memory size trends over several generations. Table 5.3 compares performance parameters for two generations of memory, DDR3 and DDR4. Maximum data rates almost double from one generation to the next. However, CAS latency, a key timing parameter, has not changed significantly and in some cases has increased. Data rate improvements have come from improvements in the architecture of memory chips, not from scaling-related improvements in the core devices and circuits.

5.4.1 **Memory structures**

While registers provide a form of memory, we generally use the term *memory* to refer to bulk memory. A semiconductor memory can store a large number of values of which only one (or a few) can be accessed at a time. The interface to a simple memory is shown in Fig. 5.19: unidirectional address lines, bidirectional data lines, and a read/write control signal.

Internally, memory is organized as a two-dimensional array as shown in Fig. 5.20. Each row and column of the memory contains a circuit that stores 1 bit of memory. The address is used to select the row and column to choose the bit of memory—we can form the address as $address\langle n - 1 : r\rangle = row$, $address\langle r - 1 : 0\rangle = column$. We use the term **access** to refer to either a read or write operation. A **random-access**

Table 5.2 Memory Capacity Trends [ITR11]					
	2011	**2014**	**2017**	**2020**	**2023**
Capacity (GBits)	68.72	137.44	137.44	549.76	549.76
Generation	64G	128G	128G	512G	512G

Table 5.3 Memory Speed for DDR3 and DDR4 [Mic11; Mic15]		
	Data Rate (MTransactions/s)	**CAS Latency (ns)**
DDR3	1066–1866	13.1–13.91
DDR4	1866–3200	13.32–15

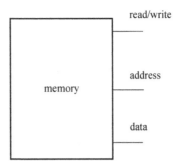

FIGURE 5.19

Interface to memory.

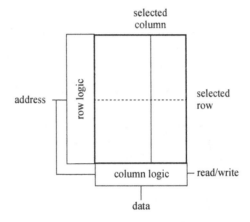

FIGURE 5.20

Organization of a memory array.

memory is one whose locations can be accessed in arbitrary patterns of addresses. Modern disk drives and early mercury delay lines, in contrast, do not allow random access.

Just as we can design registers using either dynamic or static circuits, we can design dynamic RAM (**DRAM**) or static RAM (**SRAM**). The basic principles of storage are the same as for registers, but adaptations are made for bulk memory. We will deal with flash memory separately in Section 5.5.2.

Fig. 5.21 shows the schematic for a dynamic RAM cell; this circuit was invented by Robert Dennard, who also developed ideal scaling theory [Den68]. As with the dynamic register, the bit is stored as charge on a capacitance. However, modern DRAM, known as 1-transistor or **1T DRAM**, uses a specially built capacitor rather than transistor gate capacitance to store the value. Building a specialized capacitor allows more capacitance to be built in a smaller area, increasing the number of bits that can be built

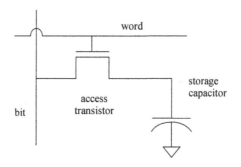

FIGURE 5.21

Circuit diagram for a DRAM cell.

on a chip. To write the DRAM cell, the row circuitry activates the **word line** to turn on the access transistor while the column circuitry sets the **bit line** to the value to be written to the cell, either a high or low voltage. Writing the DRAM requires more complex circuitry that deals more directly with analog voltages since there is no amplifier for the bit value, unlike the dynamic register. The word line is activated for a write, but the bit line is set to a high voltage. The column circuitry then senses the bit line voltage to determine how much it changes. If the storage capacitor was charged to a high voltage, the bit line will remain high. If the storage capacitor was discharged, the bit line voltage will droop, but it will not fall all the way to ground. The column circuitry must amplify the voltage on the bit line to generate a proper logic value at the DRAM output. In addition, the column circuitry must pull the bit line down to ground to ensure that the storage capacitor is discharged. The bit line voltage fell because some of the charge moved to the storage capacitor, wiping out the value stored there. The rewrite ensures that a proper value is left on the storage capacitor.

We can compare DRAM and SRAM on several characteristics as shown in Table 5.4: DRAM is denser (more bits per unit area) and uses less energy; SRAM is faster. These characteristics lead us to use DRAM and SRAM for different purposes in different parts of the computer system.

We can analyze how DRAM delay behaves under scaling using the circuit of Fig. 5.22. When the access transistor is off, the circuit consists of two capacitances: C_{bit} for the bit cell and C_{line} for the bit line:

$$V_{bit} = \frac{Q_{bit}}{C_{bit}}, \quad V_{line} = \frac{Q_{line}}{C_{line}} \tag{5.21}$$

Turning on the access transistor creates a **capacitive charge divider**. The voltages across the two capacitors (V_{bit} and V_{line}) must be equal, so the charge redistributes itself to move the line voltage from V_{line} to a common voltage V_{bl}. The capacitors are in parallel, and the amount of charge on each capacitor depends on its capacitance:

$$V_{bl} = \frac{Q_{line} + Q_{bit}}{C_{line} + C_{bit}} \tag{5.22}$$

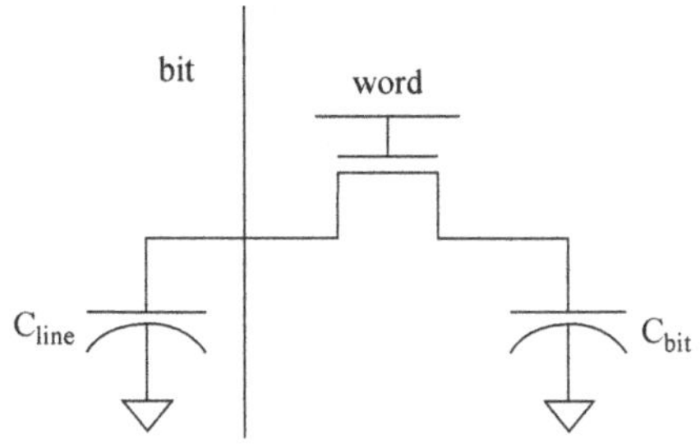

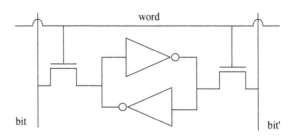

FIGURE 5.22

Model for DRAM sensing.

The worst case is when C_{bit} is discharged. $Q_{bit} = 0$, so

$$\frac{V_{bl}}{V_{line}} = \frac{C_{line}}{C_{line} + C_{bit}} \tag{5.23}$$

If we assume that scaling shrinks both capacitances at the same rate, then the best case is that DRAM does not get faster with scaling. But if chip size also scales, then the bit line will become longer, causing C_{line} to increase relative to C_{bit} and resulting in DRAM access time increasing.

The charge on the storage capacitor can leak away and the DRAM cell has no circuitry to restore its value. A bit's lifetime is typically around 1 ms at room temperature. We can **refresh** the value by reading the cell; the column circuitry naturally ensures that a proper bit value is left on the storage capacitor. DRAMs include circuitry that refreshes every bit in the memory periodically. A bit that is being refreshed cannot be accessed until the refresh operation is complete. Rather than stall all memory accesses while every bit is refreshed, the refresh logic refreshes one row at a time and waits between row refreshes.

As shown in Fig. 5.23, an SRAM cell stores its value in a pair of cross-coupled inverters, just as in the static register. However, the standard SRAM cell, known as **6T SRAM**, has two access transistors and 2-bit lines. The logical senses of the 2-bit lines are opposite: in this case, the left-hand bit line holds the true value and the

FIGURE 5.23

Circuit diagram for an SRAM cell.

Table 5.4 Comparative Characteristics of DRAM and SRAM

	DRAM	SRAM
Density	Very dense	Less dense
Performance	Slower	Faster
Energy	Less energy	More energy

right-hand side holds the false value. The word line turns on both the access transistors. To write the cell, the column logic puts the desired value of the bit on the bit lines: the true form on *bit* and the complement on *bit'*. To read the cell, the column logic puts an intermediate voltage on both bit lines and allows the cell to drive the bit lines to their logical values.

Table 5.4 compares the characteristics of DRAM and SRAM. DRAM provides more performance at lower energy levels but is slower than SRAM. The speeds of both DRAM and SRAM depend on the length of the bit line. This effect is most clearly seen in the relationship between cache size and access time—larger caches require longer access times.

5.4.2 Memory system performance

DRAM's behavior under scaling is very different from that of logic: DRAM gets denser with scaling but does not get much faster. DRAM density benefits very directly from smaller device sizes, given that it is composed of a huge number of copies of the same basic cell. (The cell capacitor is a special challenge for manufacturing, but its dimensions have shrunk considerably over time.) As a result, the number of DRAM bits per chip has doubled per generation for several decades.

However, DRAM access times have increased only by a small amount; access time has been in the range of 80 ns for many generations. Dennard scaling modeled the delay of amplifying logic gates, but DRAM cells do not contain amplification so it is reasonable to expect that ideal scaling would not directly apply. Fig. 5.24 illustrates the growth of the **memory wall** over time. The ratio of memory access time T_{mem} to clock period T is the figure of merit for the relative performance of logic and memory. In 1980, the VAX-11/780 ran at a clock rate of 10 MHz with a memory access time of 1.2 µs, with the CPU about 12 times faster than main memory. Today, with gigahertz clock rates and DRAM access times in the neighborhood of 80 ns, main memory is about 80 times slower than the clock speed of the chip. Modern processors can also execute several executions at once, increasing the demand of the CPU for memory bandwidth.

One important solution to the memory wall has been to use SRAM to build an intermediate stage in the memory system known as a **cache**. As shown in Fig. 5.25, the cache is placed between the CPU and DRAM. When the CPU accesses memory, it sends the request to the cache. Since SRAM is faster than DRAM, the cache can answer the request faster than DRAM. But because SRAM is less dense, the cache

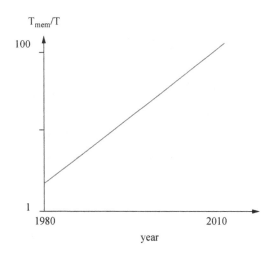

FIGURE 5.24

Growth of the memory wall over time.

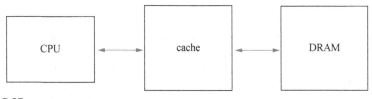

FIGURE 5.25

A memory system using both SRAM cache and DRAM.

does not hold the entire contents of memory, but only the contents of certain addresses. If the requested location is not in the cache, then the cache passes the request onto the DRAM, known as a **cache miss**. A request that can be found in the cache is known as a **cache hit**. The time required to satisfy a particular access depends on whether the location is in memory. If enough requests are to locations that are resident in the cache, then the average access time seen by the CPU is faster than for a DRAM-only memory system. The average access time of a DRAM+cache memory system is

$$t_{av} = t_{hit}p_{hit} + t_{miss}(1 - p_{hit}) \tag{5.24}$$

where t_{hit} is the access time for a hit, t_{miss} is the access time for a miss, and p_{hit} is the probability that an access refers to a location stored in the cache.

A significant component of the performance of a cache comes from **prefetching**. Rather than fetch only the word requested, the cache fetches a set of W locations around that location. If later accesses refer to those locations, then they will be in the cache unless they have been flushed by intervening accesses. Prefetching is a simple mechanism for predicting access patterns.

Hit rate depends on the behavior of the program. We can identify several common cases of execution behavior and the effect of prefetching on hit rate:

- Many array-oriented programs access locations sequentially. In an ideal case, $p_{hit} = \frac{W-1}{W}$ since the first access fetches W words, all of which are used.
- Many data structure—oriented programs chase pointers and access widely varying addresses. In the worst case, $p_{hit} = 0$.

Two important parameters characterize memory and processor performance:

- The **DRAM/cache ratio** $M = \frac{t_{miss}}{t_{hit}}$ captures the relative speeds of main memory and cache.
- The **cache/clock ratio** $C = \frac{t_{hit}}{T}$ captures the ratio of CPU and cache speed.

We can rewrite Eq. (5.24) to make use of the DRAM/cache ratio:

$$t_{av} = t_{hit}[p_{hit} + M(1 - p_{hit})] \tag{5.25}$$

At very high hit rates, DRAM speed does not matter. But at lower hit rates, the DRAM/cache ratio M becomes an important factor in performance. The importance of M helps to explain the importance of multiple levels of cache. If a fast cache is paired with DRAM, the large disparity in their speeds captured by M results in a larger average access time. Introducing several levels of cache can be modeled as a series of M values that relate adjacent cache levels. Pairing a smaller, faster cache with a larger, somewhat slower cache helps to mitigate the cache miss penalty.

The cache/clock ratio is a further penalty on performance scaling on the top of DRAM/cache ratio. To scale program performance, we must scale all three parameters: CPU speed, cache speed, and main memory speed.

SRAM access times depend on the length of the word line, so larger memories are slower. Many modern caches have several levels of cache: level 1 cache is closest to the CPU and is both small and fast; level 2 cache is larger and slower; and additional levels may be used.

5.4.3 DRAM systems

Pinout is a major bottleneck in memory systems. CPU performance is limited by memory access time and bandwidth. Pin impedance is an important limit on the speed of the memory—processor interface.

DRAM packaging is designed to minimize manufacturing cost by minimizing the number of pins required. DRAMs traditionally divide the address into two components, row and column. This decision allows address pins to be reused. It also means that addresses must be multiplexed. However, this potential limitation is used to advantage with page-mode addressing schemes.

The basic DRAM circuit is asynchronous and does not make use of clocks. Early DRAM chips were asynchronous and had no clock pin. Modern **synchronous DRAM** uses a clock to pipeline memory accesses and improve throughput.

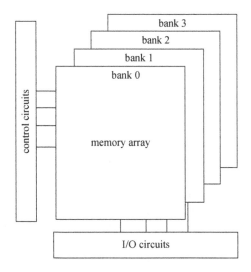

FIGURE 5.26

DRAM organized into banks.

Building wide DRAMs that provide many bits for a single access takes advantage of the fact that many accesses in programs are to successive addresses. A wide access composed of several words can be stored in the cache and then delivered to the processor as requested. Modern DRAMs are also organized internally as multiple **banks**, not a single large array, as shown in Fig. 5.26. Multiple banks allow the memory's interface logic to access each bank in parallel using a separate address. The DRAM still has one set of pins for addresses and data that are shared among the banks. The memory control logic is able to start one access on one bank and then initiate a separate access in another bank, overlapping the execution of the accesses.

3D packaging uses specialized structures to create vertical connections to complement the horizontal connections on-chip. Fabricating a new crystalline substrate on the top of an existing chip is difficult so these technologies combine several chips together using specialized vias [Loh07]. Chips can either be combined front-to-front or back-to-front. The front-to-front method allows only two chips to be combined while the back-to-front method allows many chips to be stacked.

Back-to-front stacking relies on **through-silicon vias** (**TSVs**) for vertical connections. Vias are cut through the substrate and filled with metal to provide connections on the back of the substrate. The TSVs must be large enough to make reliable connections with the pads on the opposite chip. While these vias have higher parasitic values than on-chip wires, their parasitic impedance is considerably smaller than that of pins.

Memory is a prime candidate for 3D integration because of the critical nature of memory delays and the relatively small number of pins required. The HBM2 standard [JED16] describes the design of stacked DRAM modules. It provides up to eight channels per stack with 128 pins per channel running at 1 Gb/s.

5.4.4 **DRAM reliability**

The capacitor used to store charge in a DRAM cell is very small, typically 30 to 40 fF, resulting in a small number of electrons used to store a value:

$$Q_s = C_s V_s = 30 \times 10^{-15} \text{ F} \times 1 \text{ V} = 30 \times 10^{-15} \text{ C} \tag{5.26}$$

$$n_c = \frac{Q_s}{q} = 187,500 \text{ electrons} \tag{5.27}$$

Because the value is not restored with amplifiers, stray charge can change the value of a cell. Alpha particles are an important cause of DRAM errors [May79]. An alpha particle is an ionized helium atom that is missing its electrons. When an alpha particle enters silicon, it decelerates. The energy of its deceleration ionizes silicon atoms, leaving a trail of electrons that can contaminate the device. Elevating an electron in silicon to the conductive band requires 3.6 eV; the alpha particle loses energy when it enters silicon at a rate of 150 keV/μm. Because the alpha particle has so much energy, a single particle can release hundreds of thousands or even millions of electrons. The electron—hole pairs are generated within 2—3 μm of the surface of the silicon. That charge can then be captured by on-chip structures and is more than enough to change the value of a bit in many devices. The electrons do not have to fully charge a storage capacitor to change its logical value. The amount of charge required to change the value is known as Q_{crit}; the exact amount of charge required depends on many details of the design of the cell but is a fraction of the total number of electrons that the capacitor can store.

Error correction codes (**ECCs**) can be used to detect or correct errors. Fig. 5.27 shows a simple example of an ECC. In this case, a single bit of data is represented by

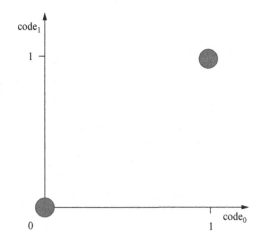

FIGURE 5.27

An error detection code.

2 bits in the code. Two of the possible four codes represent valid data; the other two codes represent errors.

5.5 Mass storage

Computer systems make use of **mass storage** to store large amounts of data. This form of storage is also known as **nonvolatile storage** because it maintains its values without requiring a power supply. This storage is organized into files by the **file system**. It may also be used by **virtual memory** mechanisms. Several very different physical mechanisms are used for secondary storage that provide different trade-offs among density, speed, and cost.

5.5.1 Magnetic disk drives

The **magnetic disk** was invented in the mid-1950s at IBM and is still very widely used in computer systems. The first disk drive, the IBM 350 Disk File, was the size of a small dish washer, stored 3.75 MB, and had an average access time of just under 1 s.

As shown in Fig. 5.28, data are stored on the disk **platter** in concentric circles known as **tracks**. Each track is divided into **sectors** that contain data, error correction codes, and synchronization and control information. A single disk has two **surfaces** that can be used to store data. Several platters are stacked together on a single **spindle**. The set of tracks at the same position on all the platters is known as a **cylinder**. Data are read by a **disk head** that is moved radially from one track to another. Each surface has its own head, allowing high data rates from parallel reads of several surfaces.

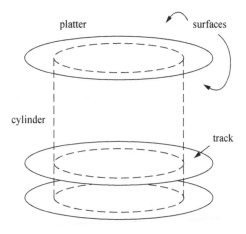

FIGURE 5.28

Organization of magnetic disk media.

Data are accessed a sector at a time. To access a sector, the **disk controller** first moves the heads to the requested track. It then waits for the appropriate sector to come under the head; each sector has an address that can be read. Once the sector is under the track, it can be read or written. Disk heads must operate extremely close to the disk to minimize the magnetic field required and the size of a bit. A typical head runs 3 nm above the disk surface. These extremely tight tolerances mean that disk drives must be kept extremely clean. A **head crash** is a collision between the head and the platter. The disk drive relies on **ground effect** to keep the head close to the surface without crashing. Compressing the air of a flying object near a surface increases the density of the air underneath. Birds use ground effect to skim over lakes. This effect means that disk drives cannot be operated in a vacuum.

The worst-case scenario for access time is:

- The heads must be moved between the opposite sides of the platter.
- The heads must wait for almost a full rotation for the desired sector to become available.

We can write a formula for disk access time as

$$T_{access} = T_{seek} + T_{rot} + T_{RW} \tag{5.28}$$

Seek time per track t_{seek} is constant, so the seek time for a distance of n tracks is

$$T_{seek} = nt_{seek} \tag{5.29}$$

The rotational time depends on the number of sectors per track s, the rotational speed r, and the number of sectors to be traversed n_r:

$$T_{rot} = \frac{1}{rs}n_r \tag{5.30}$$

T_{RW} is the time required to read or write the sector.

Example 5.5 Disk Drive Characteristics

The Hitachi Ultrastar drive [HGS12] stores data at a density of 285 Gb/in^2. Its maximum seek time is 8 ms. Its rotational speed is 7200 RPM, and its average latency is 2.99 ms. Its maximum transfer rate is 300 MB/s.

5.5.2 Flash memory

Flash memory is widely used as solid-state storage in both portable devices and servers. The basis for flash memory is the **floating-gate cell** [Kah67] shown in Fig. 5.29. The transistor gate is isolated and not electrically connected to any other circuit. A control gate is fabricated above the floating gate; the control gate is connected to driving signals as with a standard MOSFET.

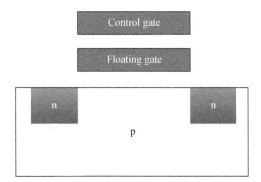

FIGURE 5.29

A floating-gate cell.

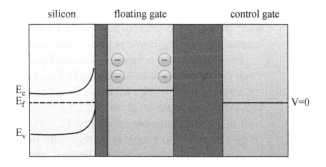

FIGURE 5.30

Band structure of a floating-gate cell with a stored charge on the floating gate [Kah67].

Fig. 5.30 shows the band structure when charge is stored on the floating gate. The effect of the floating gate charge is to increase the transistor's threshold voltage. If the threshold voltage of the transistor with no floating gate charge is V_{t_0}, the threshold with a floating gate charge $\overline{Q}$ is $V_t > V_{t_0}$. If a gate voltage V_{t_0} is applied to the control gate when no floating gate charge is present, then the transistor will turn on. If the same voltage is applied with programming charge $\overline{Q}$, then the transistor will not turn on. As a result, storing charge on the floating gate represents a 0 in the floating gate cell. The charge stored on the floating gate is very stable and can last for a long time.

The floating gate cell is operated in one of three modes [Pav99]:

- To read a cell, a positive voltage above V_{t_0} but below V_t is applied to the control gate; a positive V_{ds} is placed across the channel. If drain current flows, then no charge is stored on the floating gate and the cell stores a 0.
- To program a cell by storing charge on the floating gate, a high positive voltage is put on the control gate and a positive V_{ds} is placed across the channel.

Two different physical mechanisms can be used to apply charge to the floating gate. Fowler-Nordheim tunneling causes electrons to tunnel from the drain to the floating gate. Alternatively, hot electrons (high-energy electrons) can be used to break through the Si-SiO_2 interface and reach the floating gate. Programming time depends on the programming current. Typical programming times are in the microseconds.

- To erase a cell, the control gate is set to ground, the source to a high voltage, and the drain is left to float. Fowler-Nordheim tunneling drains the charge from the floating gate to the source.

Modern flash memory circuits do not allow individual cells in the memory to be erased independently. This erase circuitry is shared among several cells in a **block.** To change 1 bit in the block, the entire block must be erased and then all its cells must be rewritten.

The noise margins of a flash cell are determined by the ranges of voltages that represent a 0 or 1 on the floating gate. These noise margins are limited by the difference between the unprogrammed and programmed threshold voltages V_{t_0} and V_t. Increasing $V_t - V_{t_0}$ improves noise margins but also increases programming time.

Flash architectures have been extended to break the binary paradigm followed by most digital circuits. As illustrated in Fig. 5.31, 2 bits can be encoded in a single cell by dividing the $V_t - V_{t_0}$ range into four regions; this approach is known as a multilevel cell (MLC). A triple-level cell (TLC) creates eight separate regions. To represent multiple bits with reasonable reliability, $V_t - V_{t_0}$ must be increased. Access and programming circuitry also becomes significantly more complex.

Another technique used to increase flash storage density is the use of vertical structures. A vertical stack of transistors can be constructed by depositing a series of layers

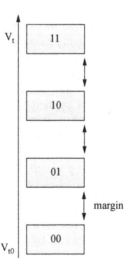

FIGURE 5.31

Voltage levels in a multilevel flash cell.

on the wafer. The combination of vertical structures and multilevel cells has resulted in huge flash storage densities.

Tunneling of charge causes physical wear that eventually causes the cell to fail. Early flash memory had lifetimes in the thousands of cycles; modern flash can withstand millions of write cycles. However, file systems tend to write much more frequently to directories than to nondirectory locations, causing the file system cells to wear out early. Flash-aware file systems move directories periodically for **wear leveling** that is designed to avoid overstressing one part of the memory.

A **solid-state disk** (**SSD**) is a flash storage device with a disk controller and interface that make it behave as a magnetic disk. Two important performance characteristics for mass storage are sustained transfer rate and latency. We can observe two general trends in magnetic disk versus SSD:

- SSD is faster than magnetic disk in both latency and transfer rate.
- Magnetic disks have symmetric read and write times while SSD writes are longer than reads.
- SSDs require sophisticated controllers to manage write-induced wear and to detect and compensate for bit failures.

Example 5.6 Magnetic Disk Versus Solid-State Disk Performance

Here are the performance metrics for two typical drives, the HGST Ultrastar A7K2000 [HGS12] and the Intel Solid-State Drive 520 [Int12]. In the case of the magnetic disk drive, we use seek time as a proxy for latency.

	Sustained Transfer Rate (MB/s)	Latency (μs)
HGST Ultrastar A7K2000	134	8200
Intel SSD 520	430 (read), 80 (write)	80 (read), 85 (write)

5.5.3 Storage and performance

Secondary storage is used for two main purposes: file systems and **demand paging**. A program can avoid file system performance issues in many cases by working in main memory. However, all programs are affected by the performance characteristics of demand paging.

Demand paging is a component of a **virtual memory** system. A program's address space is divided into **pages**. The program operates on a virtual memory image with logical addresses. Pages may not reside in physical memory at any given time; the status of program pages is maintained by a combination of hardware and software. When a program accesses a page that is not in physical memory, a situation known as a **page fault**, it is fetched from secondary storage.

We can model paging performance using a model similar to the cache model. In this case, however, segments are retrieved from the storage device. We can define a **drive/DRAM ratio** to describe the relative performance of main memory and secondary storage:

$$M_d = \frac{t_{drive}}{t_{DRAM}} \tag{5.31}$$

The average access time for a page depends on the probability of the page residing in memory P_{res}:

$$t_{page} = t_{res}[p_{res} + M_d(1 - p_{res})] \tag{5.32}$$

Example 5.6 showed that solid-state drives are about $100\times$ faster than magnetic disk drives. We can find the ratio of average paging times for SSD versus magnetic disk:

$$\frac{t_{page,SSD}}{t_{page,mag}} = \frac{p_{res} + M_{ssd}(1 - p_{res})}{p_{res} + M_{mag}(1 - p_{res})} \tag{5.33}$$

At low page residency probabilities, this ratio approaches t_{SSD}/t_{mag}.

5.6 System power consumption

Power and energy consumption are a major concern for all types of modern computer systems. We will look at power and energy at both ends of the computer system range: first servers, then mobile systems. We will conclude with an examination of power and thermal management mechanisms.

5.6.1 Server systems

High-end microprocessors consume enormous amounts of power. Even though each gate uses a small amount of energy, the huge numbers of gates results in large dynamic power consumption. Static and leakage current adds to the current requirements of large workstation and server processors.

For example, the Intel Xeon Processor E7-8800 [Int11] is specified to draw a maximum current of $I_{CC\_MAX} = 120$ A. In comparison, the Sears Craftsman Arc Welder [Sea16] is rated at 60 A. Keep in mind that the Xeon's current flows over a voltage drop of 1.3 V while the welder operates at 120 V so the welder draws more total power. Nonetheless, the current density of modern processors is impressive.

Another approach to the design of energy-efficient systems is to generalize beyond the von Neumann machine and microprocessors. **Heterogeneous architectures** provide processing elements of different types that can be tuned to particular algorithms or applications. Many **multiprocessor systems-on-chips** provide several different types of processors, including RISC processors, digital signal processors (DSPs), and hardwired accelerators. Heterogeneous architectures use less logic to perform their intended function, reducing both static and dynamic energy consumption. Many also perform these functions in less time, allowing them to run at lower clock speeds and power supply voltages.

The large power consumption of processors creates problems in the design of data centers. A single server requires not only a CPU but also DRAM, mass storage, and a network interface. Large data centers may hold tens of thousands of servers. A typical high-end data center uses tens of megawatts of electricity, 100 times more than a typical office building. The cost of electricity is typically 10% of the total cost of ownership of a data center and can exceed the cost of the hardware.

Example 5.7 Server Power Consumption

Here are typical power consumption values for the major components of a server:

Component	Power (W)
CPU	100–200
Memory	25
Disk	10–15
Board	40–50
Power/fans	30–40
Total	200–350

The CPU is the largest single power sink, but other components consume significant amounts of power.

Example 5.8 Server Power Density

A typical rack in a machine room holds 50 servers. If we assume 50 W per server, then a single rack consumes 2.5 kW, giving a monthly power consumption of 1800 kilowatt-hours (kWh) per month. For comparison, the average residence in the state of Georgia consumes 1152 kWh per month while the average Colorado residence consumes 687 kWh/month [EIA15].

Operating systems have **power management** modules that manage the power consumption of the system. Power management performs simple tasks such as using timers to put idle machines into sleep modes. They also perform more complex management tasks such as dynamic voltage and frequency scaling.

A data collection is a larger collection of server CPUs, storage, and networking. A large data center houses thousands of servers and consumes the electric energy of a town of 50,000 people. Both supplying the huge amounts of power required and dealing with the massive amounts of heat generated require careful engineering.

Example 5.9 Data Center Power Distribution

As an example of data center power distribution, consider a small data center of about 100 servers [Coe16].

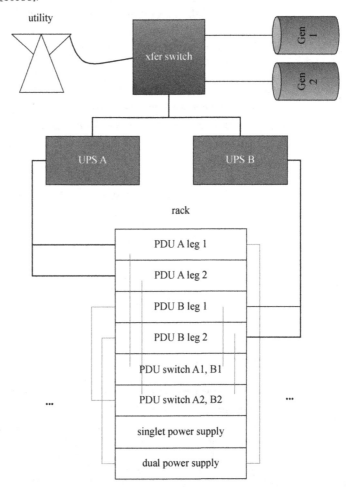

The power delivery system is designed with several goals:

- No single point of failure.
- Single-phase electrical power.

Power comes from three sources: the local utility and a pair of backup generators. The backup generators are piston engines, powered by natural gas, that are designed to automatically turn on when the utility supply fails. A transfer switch manages the switchover. However, the backup generator does not come online immediately. A pair of uninterruptable power supplies (UPSs) provide an interface between the external power sources and the servers. The UPSs provide power from their internal batteries when the external sources fail, giving continuous power to the computers and allowing the backup generator time to start. Each UPS has a capacity of 16 kVA. The two UPSs supply redundant power to each rack. Power delivery units (PDUs) take high-current inputs from the UPSs and supplies the various devices. PDU switches detect a failure of one of the UPS inputs and switch the PDU to the other supply. Critical devices have dual power supplies with inputs from two PDUs; noncritical devices have power supplies connected to only one source.

5.6.2 **Mobile systems and batteries**

Mobile systems must operate at much lower power levels than do servers that can be connected to the electric grid. The batteries that supply power for mobile systems have complex physical characteristics that must be taken into account when designing mobile systems.

Example 5.10 Cell Phone Power Requirements

The OMAP 4460 [TI] is used as a processor in cell phones. The chip includes two 32-bit CPUs, a variety of special-purpose processors and I/O devices, and on-board memory. It operates at a power supply voltage of 1.5 V and requires 2.3 A of current.

A typical TFT display with 240 × 320 resolution consumes 0.5 W.

Example 5.11 Cell Phone Power Utilization

This screen capture from a mobile phone shows the relative power consumption of its major components. Although the proportions of power consumed by the subsystems can vary, at this snapshot the screen is by far the largest power draw.

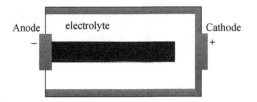

FIGURE 5.32

Structure of an electrochemical cell.

The term **battery** is commonly abused. An electrical **cell** is an electrochemical storage device, and a **battery** is a connected set of cells. Fig. 5.32 shows the structure of a typical cell. The electrical terminals are the **anode** for negative voltage and **cathode** for positive voltage. The electrolyte between the anode and cathode stores energy that is released as charge that can be used as current. We can connect cells together into batteries depending on our requirements: connecting cells in series increases the battery's voltage while connecting them in parallel increases the available current.

We can evaluate batteries using many characteristics:

- Nominal voltage and current.
- Total energy.
- Energy density.
- Shelf life.
- Number of charging cycles.
- Internal resistance.
- Performance as a function of temperature.

Table 5.5 compares the characteristics of several different types of batteries. These batteries vary widely in their voltage, capacity, usable cycles, and specific energy.

Table 5.6 compares the **specific energy**, or energy per unit weight, of a variety of materials. Different types of batteries vary widely in their specific energy. Given that the specific energy of common type of battery, lithium-ion, is about 15% the energy density of TNT, we can conclude that there is a practical limit to the energy density of batteries and that we are close to that limit. Battery specifications also increase at only a few percent per year, much more slowly than Moore's law. These limitations on the physics of batteries impose limits on the types of battery-powered systems we can design.

Table 5.5 Characteristics of Battery Chemistries [Pow16]

Chemistry	Nominal Voltage (V)	Capacity (A h)	Usable Cycles	W h/kg
Lithium ion	3.6	0.75	600	135
NiCD	1.2	1	1000	46
Alkaline	1	1.6	100	80

Table 5.6 Specific Energy of Materials [Wik16]	
Material	**Specific Energy (MJ/kg)**
U-235	1.5×10^9
Hydrogen	123
Gasoline	46
TNT	4.6
Lithium battery	1.8
Lithium-ion battery	0.72
Alkaline battery	0.67
Lead-acid battery	0.17
Electrostatic capacitor	3.6×10^{-5}

FIGURE 5.33

Typical battery discharge characteristics.

As shown in Fig. 5.33, the voltage of a battery changes over time with operation. The battery's voltage is relatively stable for a long period but as the battery comes close to exhaustion, the voltage starts to drop rapidly. The rate of change in battery voltage can be used to estimate the remaining life of the battery. Many batteries use onboard sensors and processors to provide information about the battery to the system.

Example 5.12 Smart Battery System

The Smart Battery System (SBS) standard specifies features and interfaces for batteries to provide information about their state and characteristics to a system. SBS allows the host system to monitor the battery state, including voltage, current, and temperature. A System Management Bus, based on the I^2C standard, provides a standard interface to the computer BIOS. The Smart Battery Data protocol specifies how hosts communicate with the battery. A Smart Battery Charger manages charging.

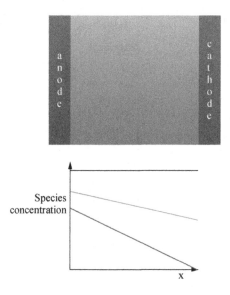

FIGURE 5.34

The Peukert Effect.

The lifetime of a battery depends on the rate at which it is discharged, known as the **Peukert Effect**. As shown in Fig. 5.34, the electrolyte is distributed between the anode and cathode. Current is drawn from the cathode, causing the **species** that supplies charge to be depleted. Not only is the total amount of species in the cell depleted, but the species near the cathode is depleted more quickly, resulting in a gradient of species as a function of distance. Once the species concentration at the cathode reaches zero, the battery ceases to function. Peukert's law gives the capacity C as a function of current I:

$$I^n = C. \tag{5.34}$$

In this case, n depends on both battery chemistry and temperature. However, if the battery is left to rest, the species diffuse back into the depleted region, reducing the gradient. As a result, we can increase battery life by occasionally reducing the current draw on the battery. Some operating systems schedule tasks to operate in bursts interspersed with rest periods to take advantage of this resting effect.

5.6.3 Power management

Modern computer systems use a combination of hardware and software to manage the power behavior of the system.

Dynamic voltage and frequency scaling (**DVFS**) is widely used to match system power consumption with required performance. DVFS is most effective when dynamic power is the dominant power consumption mode. Since power consumption varies quadratically with voltage but gate delay varies only linearly, we can improve the processor's efficiency by running at lower voltages. However, the power supply voltage must be kept high enough to provide adequate performance for the current workload. Operating system software monitors the workload, determines the proper settings for voltage and clock speed, and configures the hardware appropriately.

Race-to-dark (RTD) is designed for logic with high leakage currents. This algorithm executes tasks as fast as possible so that the processor can be put into a sleep mode that minimizes leakage current.

A simple model allows us to compare DVFS and RTD. Assume that the computing task requires n cycles of execution. The clock period is T and power supply voltage is V. Dynamic energy consumption per cycle is CV^2. The required execution time is

$$X = nT. \tag{5.35}$$

The goal of power management is to minimize energy consumption while ensuring that execution time is less than a deadline X_d. We can write the relationship between power supply voltage, gate delay, and clock period as

$$V = \frac{G}{T} \tag{5.36}$$

The energy required to execute the task is

$$E_{DVFS} = nC\frac{G^2}{T^2} \tag{5.37}$$

E_{DVFS} is minimized at the slowest execution speed that does not violate the deadline $T = X_d/n$.

RTD requires a model for leakage energy. Let the leakage energy per clock cycle be L. The total energy consumption for the task is

$$E_{RTD} = n[CV^2 + L]. \tag{5.38}$$

In this case, we minimize energy by running the system at the maximum clock frequency, corresponding to the minimum clock period $T = T_{min}$.

5.7 Heat transfer

Heat transfer, the study of how heat moves through physical objects, has become a central concern for computer system designers. Heat dissipation is a major concern for both mobile and data center systems. Mobile systems have a limited ability to dissipate heat. Although many laptops have fans, cell phones do not. Users are also sensitive to heating since they touch these devices. Cooling is a major cost in data centers. The computers must be kept within their operating temperature range, but the high heat concentrations created by large numbers of tightly packed computers create significant cooling problems.

We will start with some basic characteristics of heat transfer. Section 5.7.2 introduces basic concepts in heat transfer. Section 5.7.3 looks at the physical relationship between heat and reliability. Section 5.7.4 surveys some techniques used by systems to manage their thermal behavior.

5.7.1 Heat transfer characteristics

Heat moves through a combination of three physical mechanisms [Hal88]. **Radiation** is an electromagnetic phenomenon; heat can radiate through a vacuum. **Conduction** occurs as the result of molecular motion while **convection** depends on the bulk motion of fluids (either liquid or gaseous).

Example 5.13 Server Operating Temperatures

The Dell PowerEdge R710 [Lov16] has an operating temperature range of 50–95°F at relative humidity levels of 20–80%. It has a power-to-cool ratio (the ratio of the power consumed by the server to the power required to cool it) of 2.0%.

System-level characteristics

The operating temperature of a computer is fundamentally limited by its **maximum junction temperature** $T_{J,max}$. This temperature is measured at the transistor source/drain junctions. At high temperatures, dopants migrate sufficiently to destroy the transistors. Given that silicon is not an ideal heat conductor, the outside of the package must be kept at a considerably lower temperature to keep the on-chip devices within their operating range. A typical value of maximum junction temperature for silicon is $T_{J,max} = 85°C$. The maximum junction temperature is related to the bandgap energy of the substrate material. Silicon carbide is an example of a material with a higher bandgap that can be used to build high-temperature electronic devices.

We generally assume that all of the electrical power that goes into a chip comes out as heat. We measure heat flow in Watts, the same metric used for electric power. The most important system specification for heating is **thermal design power (TDP)**, the maximum amount of heat from the chip that its cooling system must be able to dissipate. Many modern processors cannot operate at their maximum clock rate for a significant interval without reaching their TDP limit. When they do reach the limit, they must reduce their power consumption, which requires slowing down or stopping the chip.

Physical properties

The two basic physical properties that govern convection and conduction are **specific heat** and **thermal conductivity**. Specific heat measures the relationship between heat input or output to a material and its temperature. It can be measured in units of Joules per kilogram-Kelvin. Thermal conductivity (more precisely, specific thermal conductivity) measures the relationship between temperature difference and heat flow per unit time, typically measured in Watts per meter-Kelvin.

Example 5.14 Thermal Ratings of Materials

Silicon's thermal conductivity is moderately good. Its thermal conductivity is also well matched to ceramics, which is one factor that makes ceramics a good choice for high-performance IC packaging.

Material	Specific Heat (J/kg K)	Thermal Conductivity (W/m K)	Density (kg/m³)
Silicon	710	149	2.3×10^{-3}
Aluminum nitride (fine ceramic)	740	150	3.26×10^{-3}
Carbon steel	620	41	7.85×10^{-3}

Thermal R, C

We can relate these properties of materials to the properties of specific shapes and quantities of those materials: **thermal resistance** and **thermal capacitance**. These thermal properties are analogous to their electrical counterparts, and we can use the same equations to solve for thermal behavior as for electrical behavior. We will use R and C for thermal resistance and capacitance; in most cases there will be no confusion as to whether we are referring to a thermal or electrical property. The thermal equivalents of electrical voltage and current are temperature T (measured in Kelvin, degrees Centigrade, or degrees Fahrenheit) and heat flow P (in Watts). The thermal equivalent to charge is thermal energy Q, so $P = dQ/dt$.

The thermal resistance of a piece of material is defined by a formula similar to that for the relationship between resistivity and resistance (although the thermal property is traditionally defined in terms of conductivity, not resistivity):

$$R = \frac{l}{kA} \tag{5.39}$$

where k is the thermal conductivity of the material. We can use thermal resistance to calculate the relationship between heat flow through a body and the temperature differences across the body:

$$T = PR \tag{5.40}$$

where T is the temperature difference across the body, P is the heat flow through the body, and R is its thermal resistance. This relationship is known as **Fourier's Law of Heat Conduction**.

The definition of thermal capacitance does not parallel the electrical equivalent so directly:

$$C = mC_m \tag{5.41}$$

where m is the mass of the object and C_m is the specific heat of the object's material. Thermal capacitance allows us to determine the relationship between heat flow and temperature:

$$P = C \frac{dT}{dt}. \tag{5.42}$$

This formula can be derived from **Newton's Law of Cooling**, which states that the rate of heat loss of a body is proportional to the difference in temperatures between the body and its surroundings:

$$\frac{dQ}{dt} = hA\Delta T \tag{5.43}$$

Q is the thermal energy of the body, h is the heat transfer coefficient in units of $\text{W/m}^2\text{K}$, A is the surface area, and ΔT is the temperature difference between the body and its environment.

Newton's Law of Cooling implies that bodies take time to cool down even when the heat source is removed. For a given configuration, we can combine the heat transfer coefficient and area into a single constant t_0. The solution to the law is an exponential:

$$\frac{dT}{dt} = -t_0(T(t) - T_A) \tag{5.44}$$

$$T(t) = T_A + (T(0) - T_A)e^{-t/t_0} \tag{5.45}$$

Example 5.15 Heat Transfer of Coffee

Coffee is an essential element of any computer system design. Newton's Law of Cooling implies that the rate of cooling of a cup of coffee can be slowed by lowering its initial temperature with cream. Martin Gardner, who for many years wrote the *Mathematical Games* column in *Scientific American*, observed briefly in one of his columns that coffee would cool more slowly when cream was added to it [Gar08]. Jearl Walker, author of the companion *The Amateur Scientist* column, studied this phenomenon experimentally [Walk77]. He observed that a cup of boiling water cooled smoothly to 45° in 33 min. When instant coffee was added, the cooling curve was identical for the first 15 min, then the coffee cooled faster than had the water. He estimated that the effect of radiation due to the change in color from black to white was negligible. He next added 20 ml of light cream at a temperature of 10°C to boiling water. Its temperature dropped 4° with the addition of the cream. After 5 min, the water/cream mixture followed the cooling curve of plain water. After 15 min, the cream mixture cooled faster.

5.7.2 Heat transfer modeling

Electronic systems generate heat that must be removed. Heat can directly damage the devices; reduces the lifetime of the chip; can increase the amount of leakage current;

and has many other effects. A sophisticated analysis requires both more detailed physics and finite element analysis. However, some very simple models allow us to understand the fundamentals of heat transfer for electronics.

The heat transfer characteristics of an electronic system are determined by all its components. The chip itself has thermal resistance and capacitance. The chip's package has its own thermal R and C. At high power dissipation levels, a package exposed directly to air does not dissipate heat well enough to keep the chip below the maximum junction temperature. We can use a **heat sink** to remove heat from the chip and package more efficiently. Fig. 5.35 shows the structure of one type of heat sink. The chip is located inside its package. A thermal conduction paste is placed between the package and the heat sink. Convection carries heat away from the heat sink. Fins on the heat sink increase its surface area and its convective efficiency. Forced air is often directed across the heat sink to increase convection. Some heat sinks use water to conduct heat away from a header connected to the chip and out to a radiator.

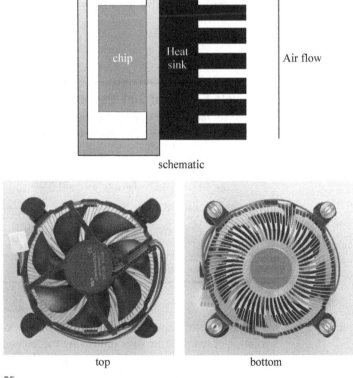

FIGURE 5.35

A chip and its heat sink.

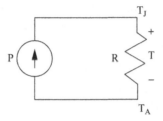

FIGURE 5.36

A thermal resistance circuit.

The specific heat capacity of water is 4.2 Joules/gram Kelvin (J/gK) while that of air is 1.0 Joules/gram Kelvin (J/gK); as a result, water can conduct the heat away much more efficiently.

Steady state   The simplest form of heat transfer analysis is steady state: we assume that the chip operates at a steady heat flow and determine the temperature of the chip. For our simple model, we will use a single value for the thermal resistance R of the package/thermal conduction paste/heat sink system as shown in Fig. 5.36. The chip junction temperature T_J and ambient temperature T_A are related by

$$T_J = T_A + PR \tag{5.46}$$

Example 5.16 Heat Sink Performance

We can see the need for heat sinks by comparing the junction temperature of a CPU with and without the heat sink. If the computer's generated power is $P = 20$ W and the ambient temperature is 20°C, we can compute the junction temperature in these two cases from the different thermal resistances. For the no heat sink case,

$$T_{none} = 20 + 20\,\text{W} \cdot 10\,\frac{°\text{C}}{\text{W}} = 220°\text{C}$$

With the heat sink,

$$T_{sink} = 20 + 20\,\text{W} \cdot 1.5\,\frac{°\text{C}}{\text{W}} = 50°\text{C}$$

The junction temperature is below the 85°C limit when the heat sink is used but well beyond that limit without the heat sink.

Transient analysis   A more sophisticated analysis of the chip's thermal behavior takes into account the evolution of temperature over time using the techniques of transient analysis. We can model the thermal behavior of a chip connected to a heat sink as a thermal RC circuit [Ska04], as shown in Fig. 5.37. The chip has its own thermal capacitance C and a thermal resistance R to the ambient.

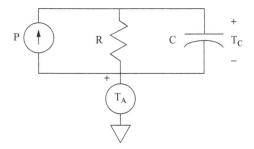

FIGURE 5.37

A thermal *RC* model of a chip.

First consider a step input from the heat source. We measure chip temperature $T_C(t)$ relative to the ambient temperature T_A; the absolute temperature of the chip is $T = T_C + T_A$. Initially, $T(0) = T_0$. The chip temperature will asymptotically approach a value HR. The temperature response of the chip is

$$T(t) = (T_0 - PR)e^{-t/RC} + PR + T_A. \tag{5.47}$$

As with electrical circuits, the thermal time constant is $\tau = RC$. The time constants for chip-level thermal behavior are typically on the order of milliseconds to seconds.

This result illustrates the utility of our earlier heat sink analysis that only considered thermal resistance. At steady state, the temperature of the chip is PR. We need to choose a thermal resistance to thermal ground such that this steady-state temperature is below the chip's critical temperature. Thermal capacitance determines how quickly the chip's temperature changes.

Example 5.17 Thermal *RC* Model of Chip Temperature

We can use the thermal circuit of Fig. 5.37 to compute the thermal behavior of a CPU based on these values:

R	0.5 K/W
C	0.03 J/K
P	50 W
T_0	0K
T_A	300K

The chip temperature as a function of time is:

$$T(t) = (0 - 25)e^{-t/\left(0.5\,\frac{K}{W}\right)\left(0.03\,\frac{J}{K}\right)} + 25 + 300 = 325 - 25e^{-\frac{t}{0.015}}$$

The thermal time constant is 0.015 s. The temperature approaches a steady-state value of 325K:

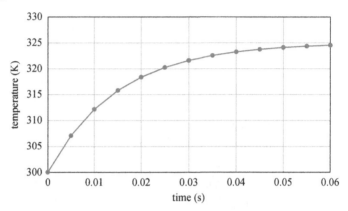

This waveform is the thermal response of the CPU to being turned on and operated at a constant power level of 50 W.

An interesting case is that of a square wave input: the chip turns on and off periodically. This case is typical of processors that are not fully loaded and do not have to run at full speed. As shown in Fig. 5.39, the chip temperature will bounce between minimum and maximum temperatures $\pm T_p$. Fig. 5.38 shows the results of a circuit simulation of a thermal square wave. In this case, $P = 50$, $R = 0.4$, $C = 0.03$. Processor temperature cycles between the ambient of 300 and 320K.

We can use a result from circuit analysis to find the steady-state temperature bounds; this analysis assumes a 50% duty cycle for CPU operation. This case is slightly easier to analyze if we assume that the heat source alternates between being heated from a heat source P and cooled by a source $-P$; this translation of the input

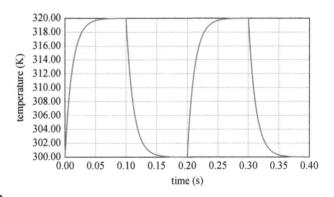

FIGURE 5.38

Thermal waveform for a thermal square wave.

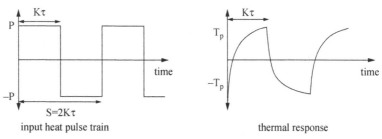

FIGURE 5.39

Peak-to-peak waveform from a thermal square wave input.

values does not affect the result. Let us assume that the square wave has a period S and 50% duty cycle (on for half the period); we will use $S = 2K\tau$. We can write the upward and downward temperature waveforms (relative to time t_0) as

$$T^u(t + t_0) = (-T_p - P)e^{-t/RC} + P \qquad (5.48)$$

$$T^d(t + t_0) = (T_p + P)e^{-t/RC} - P \qquad (5.49)$$

We know that the upward waveform's value at $t = K\tau$ is T_p and the downward waveform's value at $t = K\tau$ is $-T_p$. Substituting into Eqs. (5.48) and (5.49),

$$T_p = (-T_p - P)e^{-K} + P \qquad (5.50)$$

$$T_p = (T_p + P)e^{-K} - P \qquad (5.51)$$

This gives

$$\frac{T_p}{P} = \frac{1 - e^{-K}}{1 + e^{-K}} \qquad (5.52)$$

An ideal heat sink is harder to construct than is an ideal ground in an electrical circuit. When building a circuit, we can easily connect the ground terminal to an earth ground using a low-resistance wire. Building a low thermal resistance connection to an effectively infinite reservoir of thermal energy requires more effort on both the connection and the heat reservoir.

Thermal properties of a chip are important not just relative to the package and ambient but also in how parts of the chip affect each other. Different subsystems on a chip have different thermal characteristics based on both their design and how they are used during execution. As a simple example, multicore processors often have one or more cores idle much of the time. Cores that are in use will heat up, but that heat will also spread to adjacent cores that are idle.

We can model heat transfer on-chip using networks of thermal RC elements. Fig. 5.40 shows a simple model for a two-core processor. Each core has its own thermal resistance and capacitance as well as its own heat source. The two cores are connected by the thermal resistance R_{12} of the boundary between them. One simple but

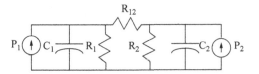

FIGURE 5.40

A thermal model for a pair of adjacent cores.

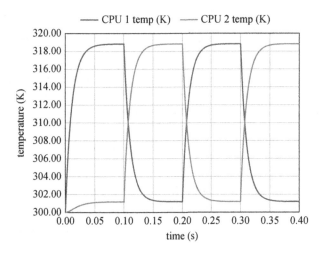

FIGURE 5.41

Thermal waveforms for a two-core system with alternating operation.

interesting use case for this two-core model is heat sources as alternating square waves: one core is on while the other is off. In this case, the on core heats up not only itself but also the other core. As a result, the off core does not cool down as quickly as it would if it were thermally isolated. Fig. 5.41 shows the results of simulating a two-core system in which the cores alternate operation, with $P = 50$, $R_1 = R_2 = 0.4$, $C_1 = C_2 = 0.03$, $R_{12} = 6$. In the first cycle, some heat spreads from the first CPU to the second. Afterward, the thermal waveforms for the two cores are complementary. The minimum temperature is somewhat above and the maximum temperature somewhat below that of the single-core case. When a CPU is idle, it absorbs some heat from the other CPU, raising its own temperature and lowering that of the other CPU.

5.7.3 Heat and reliability

Heat is an important reliability concern not just because of catastrophic failures but also because heat accelerates aging. The hotter chips run, the faster they age and degrade.

We can understand the reliability and aging effects of heat through **Arrhenius' equation**, which governs many physical processes [Sie82; Pan09]. It relates the **activation energy** E_a of the physical process to the rate r of that process:

$$r = Ae^{-E_a/kT} \tag{5.53}$$

The activation energy E_a is a basic property of the physical process that is related to the energy required to promote electrons to high orbits. The **Arrhenius prefactor** A is usually determined experimentally.

Example 5.18 Activation Energies for Failure Mechanisms

Here are some failure mechanisms and their activation energies for Arrhenius' equation [Pan09].

Oxide film defect	0.3–1.1 eV
Iconicity drift (Na ions in oxide film)	0.7–1.8 eV
Slow trap	0.8–1.2 eV
Electromigration disconnect	Aluminum: 0.5–0.7 eV Copper: 0.8–1.0 eV
Aluminum corrosion	0.7–0.9 eV

Electromigration is an example of a temperature-related failure mechanism [Bla69; Pan09]. When a wire heats, some of its atoms break their molecular bonds and become free atoms. Current flowing through the wires can interact with these free atoms and cause them to move. Current densities in the range $10^5 - 10^6$ A/cm^2 are sufficient to move significant numbers of wire atoms. This movement causes some parts of the wire thicken while others grow thinner. The thinner parts of the wires have higher resistance, causing them to heat more, thereby increasing electromigration. This cycle ultimately causes the wire to break. The mean time to failure of a wire under electromigration as a function of current density J can be modeled using Black's equation [Bla69]:

$$MTTF = AJ^{-n}e^{E_a/kT} \tag{5.54}$$

Typically, $1 \leq n \leq 3$.

High temperatures accelerate many aging mechanisms even at levels well below those that cause catastrophic failure. However, we can make use of this temperature dependence [Ska08]. We can model a chip's lifetime based on its temperature-dependent consumption of that lifetime. The life consumption rate can be modeled as

$$R(t) = \frac{1}{kT(t)}e^{-E_a/kT(t)} \tag{5.55}$$

The amount of chip life consumed depends on the temperature profile over time:

$$\varphi_{th} = \int_0^t \frac{1}{kT(t)} e^{-E_a/kT(t)} \qquad (5.56)$$

We can extend lifetime in several ways. We can design the chip to minimize hotspots; a failure at any point in the chip causes a system failure. The operating system can also spread the workload between cores to even out temperature and keep all cores aging at the same rate.

5.7.4 Thermal management

As with power management, a combination of hardware and software is used to manage the thermal behavior of the system. Thermal management is particularly challenging thanks to the long time constants of thermal behavior.

On-chip temperature sensors rely on the temperature dependence of current through a p-n junction. Although a casual inspection of the Shockley diode equation of Eq. (2.45) would suggest that diode current falls with increasing temperature, the saturation current J_0 increases exponentially with temperature due to the temperature dependence of D_p, p_{n0}, and L_p [Sze81]. This technique is used as a **bandgap reference** since the slope of the J_s versus $1/T$ depends on the energy gap E_g. A simple temperature sensor circuit can be built using a diode. A more sophisticated circuit [Lee97] uses bipolar transistors as the temperature sensors.

Modern processors take a variety of actions to protect the CPU against damaging overheating, including both direct hardware action and software notifications. The Intel Xeon e7-8800 [Int11], for example, provides two forms of direct hardware protection that are triggered when temperature sensors detect that the processor has reached its thermal limit:

- Intel Thermal Monitor 1 turns the clocks off and on at a duty cycle chosen for the processor type, typically 30–50%.
- Intel Thermal Monitor 2 uses dynamic voltage and frequency scaling mechanisms to reduce both the clock speed and power supply voltage of the processor.

TM 2 is activated first, then TM 1 is activated if necessary. Both mechanisms are required to be enabled by BIOS to ensure that the processor operates within its specified operating temperatures. The processor also implements an on-demand mode that allows software to issue commands to reduce power consumption by imposing a duty cycle on the clock. In addition, the processor provides a Platform Environment Control Interface (PECI) to allow the system to query the processor for thermal monitoring as well as power and electrical errors.

5.8 Synthesis

Based on our analysis of computer systems, we can identify some fundamental physics-based challenges to the design of computers:

- The memory wall—the failure of memory to keep up in performance with logic—is due to the basic physics of the memory mechanism.
- The power wall is due in part to the nonideal scaling of logic and in part due to leakage current.
- Clock rates have plateaued due to energy concerns. The speed of signal propagation also limits the size of synchronous system that we can create. Asynchronous interfaces must take metastability into account.
- Global interconnect incurs long delays.
- The high power density of modern systems creates significant problems in removing the waste heat generated by computers.
- Heat generation causes not just catastrophic failure but also reliability problems.

Questions

Q5-1 The clock input to a register has a capacitance of 50 fF. You are driving 100,000 registers at a rate of 1 GHz. The clock swings through a power supply voltage of 1 V. How much power is consumed driving the registers' clock input capacitance?

Q5-2 A chip with 1024 registers is driven by a buffered binary clock tree.
  a. How many inverters are in the clock tree?
  b. If an inverter in the clock tree consumes of 50 fJ for a $0 \rightarrow 1$ or $1 \rightarrow 0$ transition, how much power is consumed by the clock tree if the clock runs at a frequency of 1 GHz?

Q5-3 A clock network drives 16,384 registers. The clock input to a register connects to two transistor gates, each with $C_g = 0.9$ fF.
  a. How many stages are in the clock tree if each driver drives four loads?
  b. How much current is drawn by the clock network for a rise time of 0.2 ns to a voltage of 1 V?

Q5-3 You will analyze power/performance trade-offs for SRAM caches as we vary the number of words n in the cache. The power consumed by the cache is $P = n \cdot 0.2$ mW. The hit rate is $n \cdot 476 \times 10^{-6}$ for $0 \leq n < 2048$ and 0.98 for $n \geq 2048$. Assume $t_{hit} = 1$ ns and $t_{miss} = 100$ ns. You should plot t_{av}/P, the cache access time per unit power, for $128 \leq n \leq 4096$.

Q5-4 Plot t_{av} versus M for a memory system. Assume $t_{hit} = 5$ ns, $p_{hit} = 0.98$, $10 \leq M \leq 500$.

Q5-5   A DRAM bit line has a total capacitance of 30 fF. What value of bit cell capacitance gives $V_{bl}/V_{line} = 0.5$?

Q5-6   A DRAM bit line is of size 40 nm × 5000 nm with unit capacitance of $C = 500 \frac{fF}{\mu m^2}$. The DRAM cell has a capacitance of 40 fF. If the bit line is charged to 1 V and the cell is discharged (0 V), what is the percentage change in bit line voltage during a read?

Q5-7   If memory delay stays constant from generation to generation while memory density doubles with every generation, how does the number of banks per memory change from one generation to the next?

Q5-8   You are given performance parameters for a computer system over two technology generations:
1.   Generation n: clock speed 1 GHz, one instruction per clock cycle, 50% of instructions access memory, cache access time 1 ns, main memory access time 80 ns, hit rate 0.95.
2.   Generation $n + 1$: clock speed 1.75 GHz, one instruction per clock cycle, 50% of instructions access memory, cache access time 1 ns, main memory access time 80 ns, hit rate 0.95.
     How does the required main memory bandwidth change from generation n to $n + 1$?

Q5-9   Does Dennard scaling predict a power wall? How does Dennard predict that heat flux will scale with technology generations?

Q5-10   Plot virtual memory page access time t_{page} versus M_d. Assume $t_{pres} = 80\ ns$, $p_{pres} = 0.9$, $10 \leq M \leq 100$.

Q5-11   A disk system has a seek time of 0.1 ms per track and a rotation speed of 7200 RPM. The disk has 10,000 tracks each with 256 sectors. What is its worst-case access time? Time for one rotation 0.0083 s

Q5-12   Your data center has 10,000 servers. Each consumes 100 W at 120 V. Your copper power supply wire has a resistance per unit length of 1.7E-4 Ω/m. What is the maximum length of copper wire on the power supply to give a voltage drop of no more than 1 V?

Q5-13   Your chip requires a surge of 50 A of current over 1 ns at 1.2 V. A pin has an inductance of 3 nH. How many pins are required to limit the voltage drop during the surge to 5% of the power supply voltage?

Q5-14   The resistivity of copper at room temperature is 17E-9 Ωm. What diameter of copper wire is required to give a copper wire 10 m long a resistance of 2E-3 Ω?

Q5-15 A 50 W CPU can operate at a maximum junction temperature of 85°C. What total thermal resistance is required to safely operate the chip in an ambient temperature of:
    a. 25°C?
    b. 40°C?
    c. 30°C?

Q5-16 By what factor does the Arrhenius equation rate increase when temperature increases from 300 to 325K? Assume $E_a = 1.6 \times 10^{-19}$ J.

Q5-17 Given an Arrhenius reaction occurring at 300K with $E_a = 1.2 \times 10^{-19}$ J, at what temperature does the reaction occur at $10\times$ the original rate?

Q5-18 A CPU dissipates 75 W. The machine room ambient temperature is 20°C.
    a. At what temperature does the CPU operate with no heat sink and a package thermal resistance of 5°C/W?
    b. What thermal resistance is required for the chip to operate at 35°C?

Q5-19 A CPU burns 100 W. The maximum junction temperature is 85°C. The total heat sink thermal resistance is 0.5°C/W. Plot the maximum allowable thermal resistance for an ambient temperature ranging from 20 to 32°C.

Q5-20 A chip has a thermal resistance of 0.6 K/J s and thermal capacitance of 1.6 J/K. You give it an impulse of heat that causes its temperature to rise by 15°C above the ambient. How long does it take for the chip to cool to within 10% of the ambient temperature?

Q5-21 We can rewrite Peukert's law as $t = H\left(\dfrac{C}{IH}\right)^n$ where t is the actual discharge time, H is the rated discharge time, I is the discharge current, and C is the rated capacity. Assume $H = 2$ h, $I = 2.5$ A, $C = 4$ A h, $n = 1.3$. Find the actual discharge time and compare the actual value of ampere-hours delivered to the rating.

Input and Output

6

6.1 Introduction

Input and output devices are an important part of any computer system. Our capability to provide interesting I/O devices for computers is an important factor in driving demand for semiconductors. Multimedia was a major factor in the demand for semiconductors for over 20 years, first with digital audio, then HDTV, and ultimately for mobile multimedia. The high computing demands of multimedia required advanced server systems to deliver and produce content, high-performance clients to manipulate and display content, and huge quantities of memory. Today, the Internet-of-Things requires a new set of I/O devices for the smart nodes being integrated into physical systems.

The physics of many types of input and output devices builds upon the same physics that produced the transistor. **Electrostatics** is another major theme found in many of the most important I/O devices. The time constants for input and output are long relative to those required for high-speed digital logic. This allows slower effects such as electrostatics to be used. A third major theme is the micromachine or **microelectricalmechanical system (MEMS)**. The fabrication techniques can be used to build mechanical structures. When combined with mechanisms such as electrostatics or the piezoelectric effect, we can build machines that bridge the gap between mechanics and electronics.

Section 6.2 discusses basic principles of displays followed by Section 6.3 on image sensors. Section 6.4 considers touch and gesture input. Section 6.5 studies microphones. Section 6.6 introduces the principles of accelerometers and inertial sensors.

6.2 Displays

LEDs are not used as high-resolution displays, but their principles are relatively easy to understand; they are also closely related to photodiodes used in image sensors. One of the great coincidences of physics is that we can use the same material to both compute and manipulate light.

The Physics of Computing. http://dx.doi.org/10.1016/B978-0-12-809381-8.00006-7
Copyright © 2017 Elsevier Inc. All rights reserved.

The LED, invented in 1962 by Nick Holonyak [GEL15], is a semiconductor diode whose design has been optimized for the emission of photons [Sze81]. When an electron recombines with a hole, it must lose energy equal to the material's bandgap. At some bandgap values and electron energies, the energy released is in the form of a photon. An important measure of the efficiency of a photoemitter is its **quantum efficiency** η_q, which measures the ratio of recombination events that generate a photon to the total number of recombination events. Quantum efficiency decreases with temperature. The wavelength of the photon (ie, its color) depends on the energy of the recombination event. The human eye is sensitive in the wavelength range $\lambda = 0.4-0.75$ μm. Energies in the range of about $h\nu = 1.8-10$ eV produce visible light.

Because the wavelength of the emitted light depends on the bandgap, we must choose materials with the appropriate bandgap to generate light of the desired wavelength. $GaAs_{1-x}P_x$ is widely used for red and green LEDs. This type of material is known as a **III-V material** because it combines elements from columns III and V of the periodic table to create a material with properties similar to the traditional semiconductor materials of column IV but with different bandgaps than are achievable with materials such as silicon or germanium. Building LEDs capable of emitting blue light required the introduction of new materials such as InGaN and GaN. The inventors of blue LEDs—Isamu Akasaki, Hiroshi Amano, and Shuji Nakamura—received the Nobel Prize in Physics in 2014 for their work [Nob14].

The p-n junction of a diode provides the structure that promotes recombination. A forward-biased p-n junction has minority carriers injected across the junction. These minority carriers have the energy required for recombination that results in photon radiation.

We can build a display using an array of LEDs but only with the addition of several other elements. Each LED must be driven to a brightness appropriate to the pixel it represents. To control the brightness of pixels individually, we must be able to address the LEDs and their drivers. Some very large outdoor displays are built using arrays of LEDs assembled and wired by hand; using LEDs provides high brightness, large displays for outdoor events. But such displays are expensive to build. Mass production requires different techniques.

The **liquid crystal display,** invented by George Heilmeier [Hei68], uses the liquid crystal as a light valve. Nematic liquid crystals are in a liquid state but exhibit some of the properties of crystals; "nematic" refers to the arrangement of crystals under certain conditions, which resembles threads, as shown in Fig. 6.1. We can control the liquid crystal state using an electric field. The liquid crystal material is placed between two electrodes in a capacitor-like structure. As shown in Fig. 6.2, an electric field applied across the electrodes orients the crystals. Heilmeier's original LCD operated in a reflective mode. Light was applied to the front of the display, and the liquid crystals were backed with a black surface. With no applied field, the liquid crystal was transparent and appeared black against the background. When an electric field was applied to a pixel, the crystals reflected light and appeared white.

Modern LCDs operate in transmissive mode. As shown in Fig. 6.3, light is applied at the back of the display. A form of liquid crystal that twists with applied electric field

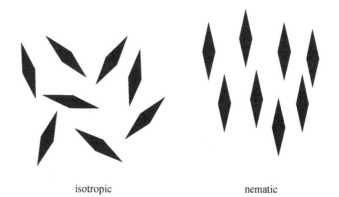

isotropic   nematic

FIGURE 6.1

Liquid crystal states.

FIGURE 6.2

Orientation of a liquid crystal to an electric field.

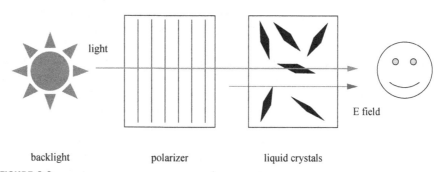

backlight   polarizer   liquid crystals

FIGURE 6.3

LCD architecture.

(*twisted nematic* liquid crystals) can be used to polarize the light. This variable polarizing in combination with a fixed polarizer can be used to control the amount of light that passes through the structure to the observer.

We still need to be able to address individual pixels and control their brightness. Early LCDs used a **passive matrix** structure as shown in Fig. 6.4. The terminals of

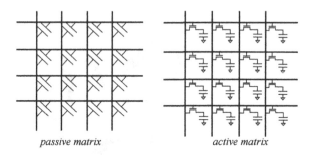

passive matrix   *active matrix*

FIGURE 6.4

Passive and active matrix liquid crystal displays.

each pixel were connected to one row and one column wire. A row is selected by activating the appropriate row wire; the column wire voltages are set to control the state of each pixel in that row. The relatively long recovery time of the liquid crystal allows us to scan the rows to build the display image. Modern LCDs use an **active matrix** architecture, also shown in Fig. 6.4. This type of display is made possible by the **thin film transistor**, which is built with thin films of material on top of a nonconducting substrate. A silicon substrate is opaque, but thin film transistors can be built on glass. The active matrix architecture is very similar to that of a memory, with the thin film transistor providing the access transistor and the liquid crystal cell playing the role of the capacitor. As with the passive matrix structure, the image is built by scanning the rows.

Liquid crystal displays provided several improvements over cathode ray tubes: they were much smaller and lighter, required lower voltages and less energy, and could be manufactured using photolithographic techniques. But the liquid crystal light valve has lower contrast than do, for example, LEDs. The light valve's off state leaks some light, causing blacks to be less intense. While LCDs have vastly improved since their introduction, the light valve approach has some basic limitations.

The **organic LED** provides a structure with an active light source that can be manufactured using standard microelectronic methods. Conductive polymers are an important class of organic semiconductors [Roy00]. Although traditional polymers are insulators, some polymers can be doped to create bandgaps that conduction by both electrons and holes. The backbone of polyacetylene, for example, provides the atomic structure required for high conductivity and electron/hole conduction. The molecular structure of a conductive polymer is much less regular than the crystalline structure of silicon, but this does not prevent it from exhibiting the same useful properties displayed by traditional semiconductors. Alan Heeger, Alan MacDiarmid, and Hideki Shirakawa shared the 2000 Nobel Prize in Chemistry for their discovery of conductive polymers [Nob00].

Organic LEDs are built much like traditional LEDs with a junction of p-type and n-type materials. They can be much cheaper to manufacture than traditional semiconductors: crystalline semiconductors are much more expensive to produce; and conductive polymers can be patterned using ink jetlike methods onto a wide variety of surfaces. Conductive polymers can also be put onto flexible materials; the animated cereal box in *Minority Report*, for example, could be built using organic LEDs. Organic LEDs produce vivid colors with good contrast. Because they are light emitters, not light valves, they can deliver both deep blacks and strong whites.

However, organic LEDs wear out much more quickly than do traditional LEDs. As an organic LED is used, its threshold voltage changes. A compensation circuit is used to adjust the drive to the OLED and maintain the proper brightness level. Fig. 6.5 shows a sample compensation circuit [Ono07]. Compensation means that the display cycle is more complex than that of an LCD; this circuit requires a four-phase scan sequence. The first step resets the state of the pixel. In the second step, the access transistor is on, and the OLED's threshold voltage is measured using a low voltage on the data line. The OLED cannot be turned on without disrupting the display, so the circuit is designed to keep the OLED's voltage just below its threshold. In the third phase, the access transistor is turned on to set the voltage of the OLED for the desired brightness. In the fourth step, the drive transistor is turned on, providing enough current to cause the OLED to emit at the desired level. The circuit is designed, so that the OLED voltage is independent of V_T during emission.

The *digital light processor* (*DLP*) is a rare example of a truly digital output device—the integration of discrete light pulses delivered by the DLP into a perceived brightness level is performed by the visual system, not by the device itself.

Fig. 6.6 shows the operation of a DLP pixel. A mirror is placed on a micromechanical pivot. Electrostatics can be used to move the mirror into either of two positions: one in which it reflects light from a source out of a lens; the other in which the light is reflected to an absorbing location within the projector chamber. The mirror's position is pulsed to control the brightness of the pixel. For example, for a display with 128

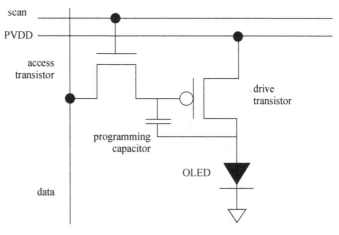

FIGURE 6.5

A compensation circuit for an OLED [Ono07].

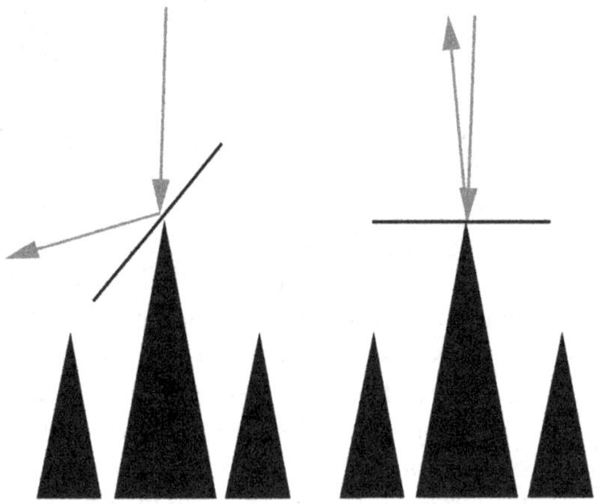

dark   bright

FIGURE 6.6

Operation of a DLP cell.

brightness levels, no pulses would result in a black pixel while 127 pulses would result in a maximum-brightness pixel. A color display can be built with three DLP arrays, each with its own red/blue/green filter. But most DLP displays use a single DLP array with a color wheel that consecutively applies red, green, and blue filtration to the fields produced by the array. The DLP display requires a lens to focus its output which puts some limits on how it can be used.

DLP provides very high contrast because it is based on a reflector, which can be made to be of very high efficiency. The use of reflectors also allows it to produce very bright displays by using a strong light source. The fact that the DLP pixel itself does not produce the light means that the high energy levels required for strong display do not have to be delivered to the pixel, which also reduces the heat load on the pixel.

Electronic ink displays also rely on electrostatics. Each pixel is represented by a ball that is white on one hemisphere and black on the other. Electrostatics can be used to flip the ball to the desired position. These displays are reflective and provide very strong contrast. But they cannot directly provide grayscale pixels. Halftoning techniques, in which collections of dots can be used to create the impression of a gray level, can be used, but at reduced resolution.

The *inkjet printer* relies on thin film transistors and micromachined nozzles to generate precise drops of ink [Nie85]. Superheating ink causes it to generate small bubbles [All85]. All the superheating bubbles are of the same size, which allows the amount of ink delivered to be precisely controlled. Superheating also causes the bubbles to move quickly, and their trajectory can be controlled by a nozzle. Fig. 6.7 shows an inkjet nozzle. Thin film techniques are used to generate a resistor

FIGURE 6.7

An inkjet nozzle.

for each nozzle; micromachining methods are used to create the nozzles. The cartridge rides on a carriage that moves over the paper at a precise rate; control circuits cause the nozzles to fire at the appropriate times.

6.3 Image sensors

The same basic photodetection mechanism is used in all image sensors. But we can build different types of image sensors that use different techniques to read the pixel values from the photodetector sites.

Photodetection uses the phenomenon of the LED in reverse: absorbing a photon promotes an electron to the conduction band [Sze81]. The responsiveness of the material to light is determined by its cutoff wavelength:

$$\lambda_c = \frac{hc}{E_g} = \frac{1.24}{E_g(eV)} \tag{6.1}$$

where h is Planck's constant. Photons of wavelength shorter than the cutoff wavelength are absorbed and create electron–hole pairs. An important metric for a photodetector is quantum efficiency η, which measures the number of carriers measured per photon.

As with LEDs, we use a junction to promote the capture of photons and generation of conduction-band electrons. The quantum efficiency of a photojunction is the ratio of electron–hole pairs generated to incident photons. The p-i-n (p-type, then intrinsic, then n-type) junction is often used for photodetection. Light absorbed by a reverse-biased junction creates electron–hole pairs that result in current flow.

Photodetectors take two common forms. A **photodiode** is a diode optimized for use as a photodetector. A **phototransistor** uses one of the p-n junctions of the transistor as the photodetector; the transistor effect then amplifies the resulting current.

As with LEDs, the composition of the material and its resulting bandgap determines the frequencies to which the photodetector is sensitive. However, photodetectors are typically used as *panchromatic* detectors that capture all frequencies of light. The fact that they are less sensitive to some frequencies is taken into account in other ways.

As shown in Fig. 6.8, a pixel in an image sensor contains several elements in addition to the photodetector. The pixel circuitry provides access to the pixel value. As a result, not all of the surface of the image sensor can be used to detect photons. The **fill factor** is the ratio of the photodetector area to the total pixel area. One way to compensate for the limited fill factor of a pixel is by using a lens to concentrate as much light as possible onto the photodetector. The microlenses must be fabricated with material of good optical quality, and their optical properties must be evenly matched across the array.

Although different photodiode materials can be used to sense light of different wavelengths, building an array of red, green, and blue photodiodes of different materials on the same chip is impractical given the small sizes required for the pixel. Instead, color filters are placed over each pixel as shown in Fig. 6.8. The filter material itself is relatively simple to handle compared with other microelectronic materials, but each pixel in the sensor array must have its own color. The most common pattern for filters is known as the Bayer pattern [Bay75] shown in Fig. 6.9. It is a 2×2 pattern with two green, one blue, and one red pixel. Two greens were chosen because the human visual system is most sensitive to green; the pair of green pixels can be used as a simple form of luminance signal.

The **charge-coupled device (CCD)** was the first successful semiconductor image sensor. Willard Boyle and George Smith received the 2009 Nobel Prize in Physics for their invention of the CCD [Nob09]. (They shared the prize with Charles Kao for his work on fiber optic communication.)

The CCD is based on the MOS capacitor [Seq75]. The capacitance of an MOS transistor depends on the applied voltage as well as the parallel plate capacitance. The interface potential for an applied voltage V_G is

$$\varphi_s = V_G' + V_0 - \sqrt{2V_G'V_0 + V_0^2}, \tag{6.2}$$

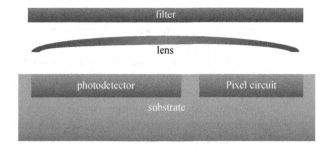

FIGURE 6.8

Cross section of a pixel.

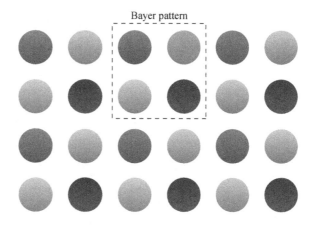

FIGURE 6.9

A filter array and the Bayer pattern.

$$V'_G = (V_G - V_{FB}) + \frac{Q_S}{C_{ox}},\tag{6.3}$$

$$V_0 = \frac{qN_A\varepsilon_0\varepsilon_{si}}{C_{ox}^2}.\tag{6.4}$$

In these equations, V_{FB} is the flatband voltage of the MOS capacitor. The total capacitance as a function of applied voltage is

$$C_{GB} = C_{ox}\frac{1}{1 + \sqrt{2\varphi_x/V_0}}.\tag{6.5}$$

We can change the capacitance of the MOS capacitor structure by applying a voltage.

We use a string of MOS capacitors to build the CCD array. Charge is moved from one capacitance to the next to form an analog shift register that is often called a **bucket brigade**. The standard way to visualize the operation of a CCD is to show each capacitor's potential well with the charge sitting at the bottom of the well. Although the charge is actually at the surface of the MOS capacitor, the potential well imagery helps us to visualize the bucket brigade behavior. MOS capacitors can be arranged in several different ways to form bucket brigades; Fig. 6.10 shows the operation of a three-phase CCD. Charge is successively transferred from one device to the next by applying voltages to each MOS capacitor. Charge will flow from one device to the adjacent device if that device's potential well is lower. A cell consists of three devices. The three phases of a clock are applied to the devices to move charge from one device to the next by manipulating their potential wells. At the end of three phases, each sample has moved by one cell.

CCDs are extremely efficient at transferring charge which means that they introduce very little noise into the image. CCDs are still used, particularly for applications

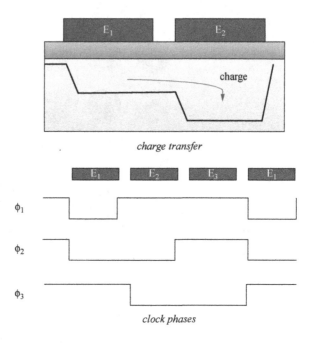

charge transfer

clock phases

FIGURE 6.10

Operation of a three-phase CCD.

that require operation in low light such as astronomy. But CCDs require specialized manufacturing processes.

The **CMOS imager**, also known as an **active pixel sensor (APS)** [Fos95; Men97], is widely used because it provides good image quality while being compatible with standard CMOS fabrication technologies. CMOS imager manufacturing is often adjusted somewhat to provide better characteristics for the photosensor, but the basic manufacturing process is shared with CMOS.

The schematic for one form of the APS cell is shown in Fig. 6.11 [ElG05]. A **photogate** is used as the photosensor. This form uses a structure known as a *pinned photodiode* due to the additional layer of doping on top to control the pinned states at the surface. A transfer gate controls access to the charge on the photogate. A pair of transistors are used to amplify the pixel value to the bit line: the bottom transistor is on when the row select line is high, allowing the top transistor to amplify the photogate output onto the bit line. The charge produced by the photogate is accumulated on the gate of the output transistor; the value of the pixel is determined by the integral of the illumination of the photodiode during the image exposure. When the reset line is high, the reset transistor reverses the bias of the photogate through the transfer gate and resets the photogate's value.

The form of the CMOS imager array is shown in Fig. 6.12. Once again, the structure is similar to that of a memory, although the values read out are continuous, not

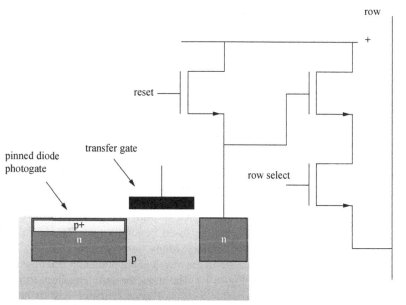

FIGURE 6.11

Schematic of an active pixel sensor (APS) cell.

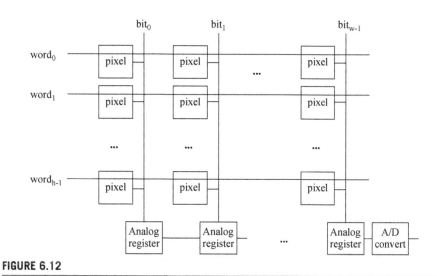

FIGURE 6.12

Architecture of a CMOS image sensor.

discrete. Given a selected word line, the bit lines can be used to read the pixel at each column. An analog shift register holds the pixel value and shifts it to an analog/digital converter.

6.4 Touch sensors

Gesture control predates the computer by many years. The first electronic musical instrument, the theremin, is controlled entirely by gestures—the musician never touches the instrument. The theremin senses the changes in capacitance between metal rod and the musician's hand. Two metal rods provide the musician with two modes of input: pitch and volume. The changes in capacitance are used to control the instrument's oscillator and amplifier. The theremin is still in use today, for example, in the theme to *Star Trek*. The capacitive sensing technique invented by Theremin is widely used beyond music. Theremin himself built a door alarm for Alcatraz Penitentiary. Many user interfaces use capacitive touch [Cyp15].

To understand the physical phenomenon and how it can be harnessed, consider the circuit of Fig. 6.13. The inductor and capacitors form a tank circuit whose resonant frequency is $1/2\pi\sqrt{LC}$. The capacitance is split into two capacitors to allow it to be connected to an amplifier that sustains the resonance. The capacitor C_2 is in parallel with the capacitance of the antenna; we lump together in this term the musician's hand capacitance C_a and any other stray capacitances in the environment. The antenna capacitance depends on the position of the musician's hand. The resonant frequency of the circuit is:

$$f_0 = \frac{\sqrt{\frac{1}{C_1} + \frac{1}{C_2} + \frac{1}{C_{ant}}}}{2\pi\sqrt{L}} \qquad (6.6)$$

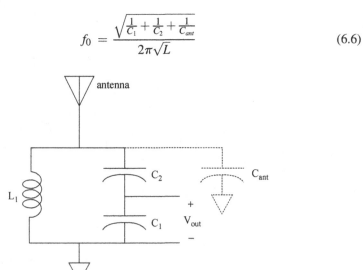

FIGURE 6.13

Circuit for capacitive sensing.

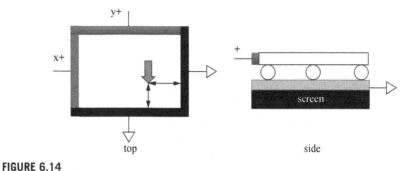

FIGURE 6.14

Architecture of a resistive touch screen.

The musician's hand forms a ground plane for the antenna that modifies the antenna's properties. Skeldon et al. [Ske98] approximated the change in capacitance ΔC_a provided by the musician's hand as

$$\Delta C_a \approx \frac{\pi \varepsilon_0 h}{10 \log\left(\frac{4x}{d}\right)} \tag{6.7}$$

where h and d are the length and diameter of the antenna, and x is the distance from the antenna to the hand.

Early touch screens for computers were based on resistance—they were two-dimensional ohmmeters. Fig. 6.14 shows the top and side views of a resistive touch screen. Viewed from the side, the touch screen has two conductive membranes separated by spacers. Both the membranes have to be clear, so that the screen can be seen through them. The bottom membrane is connected to ground. As seen in the top view, two sides of the top membrane are connected to a voltage source while the other two sides are connected to ground. The device alternates between x and y measurements. In either direction, pressing the touch screen pushes together the two membranes and makes electrical contact. The resistance at that point can be measured from which the position of the touch can be determined. Switching between horizontal and vertical measurements limits the speed with which the touch location can be measured. The relatively slow speed of touch screens makes them less useful for gesture recognition.

Most modern touch screens are capacitive. An array of capacitors is formed by wires deposited on layers of material separated by a dielectric. An electric field is applied to the capacitors; the capacitance of a finger modulates the electric field. The resulting change in charge can be sensed by addressing the capacitors.

6.5 Microphones

Audio devices have generally escaped the trend for miniaturization, which is not surprising given the wavelengths of audio signals. Both detection and generation of signals are often performed with devices whose physical dimensions are comparable to those of the signal wavelength. Audio signals in air have wavelengths in the range of hundredths of a meter to meters. In contrast, the submicron wavelengths of light are much closer to the dimensions of microelectronic devices.

All three of the three basic electrical phenomena—resistance, capacitance, and inductance—have been used to capture sound. Resistive microphones were first. The carbon microphone has a complicated history, but at least one version was invented by Edison [Edi79; Edi82A; Edi82B]. It used a pair of carbon buttons, one of which was in contact with a diaphragm. Audio waves hitting the diaphragm caused it to change the pressure on the carbon buttons, thereby changing their resistance. This microphone was cheap and effective enough to be in common use for a century.

Inductive microphones are known as *dynamic microphones*. In this case, the diaphragm is connected to an induction coil that moves relative to a magnet, inducing a current.

Capacitive microphones are known as *condenser microphones* after the traditional name for a capacitor. An interesting case is the **electret microphone** invented by Gerhard Sessler and Jim West in 1962 [Ses64]. The term *electret* is a play on magnet; the electret material is a ferroelectric material that is permanently charged. The electret material forms one plate of a capacitor that also serves as the diaphragm. Because the electret material is permanently charged, a charge does not have to be externally applied as is necessary in other types of condenser microphones.

6.6 Accelerometers and inertial sensors

An **accelerometer** is a MEMS *inertial sensor*. The inertia of a mass can be used to measure motion. Inertial sensors are used for many purposes: sports analysis devices use accelerometers to measure the motion of the body; cameras use accelerometers to measure the motion of the camera and apply image stabilization corrections.

Fig. 6.15 shows the basic design of an inertial sensor [Sch 13]. The **proof mass** is a mass of a known value. It is connected to a frame by a spring. As the frame moves, the inertia of the proof mass will cause it to resist the motion, which is transmitted from the frame by the spring. The behavior of the spring is governed by Hooke's Law [Hal88]:

$$F = -kx \qquad (6.8)$$

where x is the displacement of the spring and k is the **spring constant**. The spring and proof mass system together can be described as a second-order differential equation.

The motion of the proof mass is measured electrostatically using the proof mass as one plate in a capacitor. Electrostatics allows us to measure and exert mechanical forces because electrons exert forces on each other. Consider the case of two plates of a capacitor [Fey10B]. The work required to move an electron through a distance Δx in an electric field E is

$$\Delta W = E \Delta x. \qquad (6.9)$$

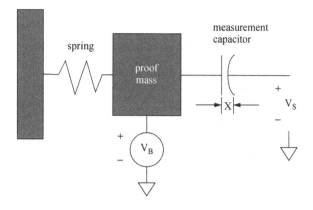

FIGURE 6.15

An inertial sensor.

The electric field between two sheets of equal and opposite charge is

$$E = \frac{\sigma}{\varepsilon} \tag{6.10}$$

where σ is the density of charge on the sheet per unit area. If the charge on the capacitance does not change as the plates move, then the change in energy of the capacitor is

$$\Delta U = \frac{1}{2}Q^2\Delta\left(\frac{1}{C}\right) \tag{6.11}$$

The total charge on the capacitor is $Q = \sigma A$ for a capacitor of area A. Substituting Eq. (6.9) into Eq. (6.11) gives

$$F\Delta x = \frac{Q^2}{2}\Delta\left(\frac{1}{C}\right) \tag{6.12}$$

The incremental change in the capacitor's value is

$$\Delta\left(\frac{1}{C}\right) = \frac{\Delta x}{\varepsilon A} \tag{6.13}$$

Substituting this formula into Eq. (6.12) gives

$$F = \frac{Q^2}{2\varepsilon A} \tag{6.14}$$

As a result of this force, the effective spring constant is lower than would be the case for the spring in isolation.

The measurement capacitor itself generates an attractive force that counteracts the spring. This effect must be taken into account in the measurements. The spring force can also be measured using the piezoelectric effect.

Similar principles can be used to build more complex motion sensors such as gyroscopes. Charge can also be applied to a capacitor to apply a force to the proof mass, which is useful in certain types of measurements.

6.7 Synthesis

- Electrostatics can be used to convert between mechanical forces and electrical signals.
- MEMS devices can be built to manipulate and measure physical phenomena ranging from movement to light.
- Image input and output depend on the fact that semiconductors can absorb and emit photons.
- Resistance, capacitance, and inductance are all used as input or output mechanisms in I/O devices.

Questions

Q6-1   The plates of a capacitor are 1×1 mm and are spaced 1 nm apart with an air dielectric of permittivity $\varepsilon = 1.0\varepsilon_0$. How much force is exerted between the plates for a charge of 1 nC?

Q6-2   The plates of an electret microphone are 3×3 mm. The dielectric between the plates is air with a dielectric constant of $\varepsilon_a = 1.0\varepsilon_0$. Plot the capacitance of the microphone for plate spacings over the range $0.1-1$ nm.

Q6-3   In the circuit of Fig. 6.13, let $L_1 = 1$ mH, $C_1 = 100$ pF, and $C_2 = 33$ pF. The antenna capacitance without a musician is $C_{ant} = 50$ pF. What range of change to the antenna capacitance is required to produce a pitch range of 3 octaves ($8\times$)?

Emerging Technologies

7.1 Introduction

We saw at the beginning of this book that the technologies that underlie today's computers were neither obvious nor foregone conclusions. Today's CMOS technology is much more advanced and complicated than the relatively simple technologies of the 1980s. The characteristics of CMOS have allowed it to dominate integrated circuits for 30 years.

CMOS has been one of the most widely used and influential technologies in history. CMOS became ubiquitous for several reasons:

- The first and most important reason for adoption of CMOS was its very low power consumption. Classic CMOS had very small static power consumption, enabling new types of portable devices. Since CMOS can operate over a wide range of power supply voltages, it was easily adapted to many operating environments. Over time, the power advantages of CMOS also translated into power savings approaches such as DVFS as well as lower thermal dissipation.
- CMOS circuits are relatively easy to design. The high input impedance of CMOS gates means that gates are only loosely coupled. Changes to one gate can slow down nearby gates but are less likely to cause complete functional failures. The structured, tub-oriented layout of CMOS circuits is also amenable to standard cell layout. As a result, transistor sizing and placement-and-routing are effective design methodologies.

CMOS is unlikely to be completely replaced any time soon. However, it does face significant challenges: high power consumption and thermal dissipation, complex manufacturing processes, limits to efficiency. The eventual end of Moore's Law has encouraged many to look for radically new alternatives. In the rest of this chapter, we will briefly consider two: carbon nanotubes and quantum computers.

7.2 Carbon nanotubes

A carbon nanotube is an example of a **fullerene**, a carbon structure in a geometric form such as a ball or a sheet. In the case of the nanotube, the carbon atoms form

The Physics of Computing. http://dx.doi.org/10.1016/B978-0-12-809381-8.00007-9
Copyright © 2017 Elsevier Inc. All rights reserved.

221

into hexagonal patterns that wrap around to form a tube. Nanotubes can, depending on the details of their composition, behave either as metallic or semiconductors.

Carbon nanotubes can be built using a geometric configuration very close to that of a traditional MOSFET [Bac01]. As shown in Fig. 7.1, a metal wire is placed perpendicular to the nanotube with a silicon dioxide layer in between. The metal wire acts as a gate while the two ends of the nanotube form the source and drain. The nanotube can be doped to control its conduction properties.

7.2.1 **Nanotube transistors**

Nanotube transistors have larger gains than do silicon MOS transistors of equivalent size. Their subthreshold conduction slope S is lower than that of a silicon transistor. This means that a smaller change in gate voltage is required to produce a given increase in channel current.

The physical mechanisms behind nanotube transistors are not quite the same as those for MOSFETs, however. Because the carbon atoms are only one layer thick, the conductive region of the nanotube device is one dimensional. This gives the electrons much less opportunity to scatter and diffuse as compared the two-dimensional motion of electrons in a MOSFET channel. As a result, electron motion is at least partially **ballistic**—the electrons travel long distances without collisions.

In addition, the control of source/drain current is not performed in the channel body but at the junctions at the end of the channel [Avo03]. The source/drain contacts create Schottky barriers between the nanotube and metal source/drain contacts. A Schottky barrier can be formed at metal—semiconductor junctions; the energy barrier creates a diode effect.

Several devices can be created on a single nanotube. By properly connecting the devices, circuits such as an inverter have been built [Bac01].

Nanotube logic structures   However, larger machines must be built out of many nanotubes with devices that are wired together. The processes used to grow nanotubes do not lend themselves to organizing the nanotubes into patterns, so separate processes must be used to organize the nanotubes. (The nanotube fabrication process also produces both metallic and semiconductor tubes, so steps must be performed to get rid of the metallic nanotubes in areas where semiconductors are desired.)

One technique for building logic networks from nanotubes makes use of a different type of nanotube structure shown in Fig. 7.2. One nanotube is suspended

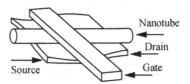

FIGURE 7.1

A nanotube transistor.

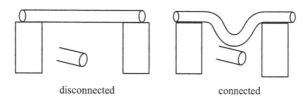

disconnected   connected

FIGURE 7.2

A switched connection built from a suspended nanotube.

over another using support structures. The two nanotubes are perpendicular to each other. By applying a voltage to the tubes, we can cause the suspended nanotube to bend closer to the bottom nanotube. Although they do not touch, tunneling allows current to flow. The programming can be changed by reapplying voltages to change their charge. Two-dimensional arrays of these suspended nanotube devices can be built; nanotube inverters can also be added to amplify signals. The two-dimensional arrays can be used to build **programmable logic arrays (PLAs)** similar to those built with MOSFETs. PLAs form an AND-OR-INVERT structure: one array performs an AND step, another array performs an OR function, and a set of inverters complete the structure to provide a full set of Boolean functions. However, at the time of this work, the nanotubes could not be arranged precisely and the devices themselves may be defective, so additional wires and devices are included in the structure to provide spares.

More recently, a new technique has been used to arrange nanotubes into well-organized arrays [Par12]. A substrate is coated with a chemical to which the nanotubes will adhere. Electron beam lithography is used to create silicon dioxide islands that cover parts of the surface. When nanotubes suspended in a solution are put on top of the surface, they selectively adhere in a pattern determined by the lithography. The substrate, which is made of highly doped silicon, is used as a backgate for the devices.

Nanotube computer   Shulaker et al. reported the first carbon nanotube computer in 2013 [Shu13]. Their computer consisted of 178 carbon nanotube FETs, each built from between 10 and 200 carbon nanotubes. The von Neumann memory was off-chip. Their computer executed a single instruction, subtract and branch if negative; this instruction is Turing complete and can be used to write arbitrary programs. They implemented a non-preemptive multitasking operating system.

7.3 Quantum computers

Quantum computers have reached the stage of experimentation but not regular use. The term *quantum computing* refers, at this point in history, not to a single technology but to several different types of machines with different uses; their commonality lies in their reliance on quantum phenomena. Interest in quantum computers was originally driven by theoretical results on their extreme efficiency thanks to the reversibility of their computations. More recent interest has focused on other useful properties: how quantum states can be used to quickly search large state spaces, and on the uses of quantum entanglement for secure communication.

Reversible computation   We have implicitly assumed throughout this book that computing is irreversible. Our analysis in Section 3.7 of error rates relied upon this assumption. But computers do not have to operate irreversibly, and reversible computation is very efficient. We have precedence for the usefulness of reversible physical operations from thermodynamics. The Carnot cycle describes an idealized reversible heat engine. Real engines cannot perform exactly as the Carnot cycle requires, but reversible engines are very efficient.

Landauer's original argument [Lan61] on the minimum energy analyzed the energy required to perform an irreversible operation. The operation is made irreversible by erasing the old result. Once the old result has been erased, the operation cannot be reversed. However, Bennett showed that we can reformulate computing operations so that they do not require erasing [Ben73].

First, consider the mathematical formulation of reversible operations. A function f is a map from a domain X to a range Y:

$$f : X \rightarrow Y \tag{7.1}$$

However, not all maps qualify as functions. To be a function, there must be at most one value in the range for each value in the domain. If the function is drawn as a graph, a point on the X axis will be represented by at most one point on the Y axis—the function curve will not double back on itself. The inverse of f is

$$\tilde{f} = Y \rightarrow X \tag{7.2}$$

For $\tilde{f}$ to be a function, each value in its range X must correspond to at most one value in X. In this case, for any x, $\tilde{f}(f(x)) = x$ is uniquely defined since there is only one possible result at each mapping.

Many Boolean functions are not reversible. For example, $OR(a,b) = 1$ could result from three different values of its inputs: $a = 1$, $b = 0$; $a = 0$, $b = 1$; $a = 1$, $b = 1$. Fig. 7.3 is an example of the exchange operator. This operation is reversible—applying it twice to any given pair of inputs results in the original input values being returned. We can in general make reversible versions of irreversible Boolean operations by adding outputs that allow us to reconstruct the original inputs.

Bennett showed how to define a reversible Turing machine that could be run forward and backward interchangeably. The most direct way to build a reversible

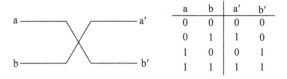

a	b	a′	b′
0	0	0	0
0	1	1	0
1	0	0	1
1	1	1	1

FIGURE 7.3

An exchange operator.

machine would be to save all its intermediate states, but this would require a great deal of storage for the final output, much of which would be of no use. His machine uses three tapes: one is used to temporarily record the history of the computation to allow reversibility; one holds the final output; the third holds the reconstructed input. To be sure that each operation on each head of the Turing machine is reversible, operations are designed so that read-write-shift is not allowed in a single step but instead is broken into separate read-write and shift operations.

DNA reversible computers   Bennett pointed out that chemical processes provide realistic examples of efficient reversible processes that perform operations similar to computing [Ben73]. He pointed out that many chemical processes related to DNA and RNA are reversible and that the manipulations performed on DNA and RNA are both complex and operate sequentially.

Adelman demonstrated the use of DNA for combinatorial optimization [Ade94]. He encoded the structure of a small graph in DNA [Ade94]. Each vertex in the graph was represented as a DNA sequence, and edges in the graph were represented as sequences that encoded the source and sink endpoints of the edge. A sequence of chemical reactions were performed iteratively to find the directed Hamiltonian path through the graph.

Quantum reversible computers   Quantum mechanics provides a set of operations that can be used to build a reversible computer [Fey85]. The state of a quantum mechanical system can be described in terms of a set of **base states**. The transition from one state to another is described as a linear combination of transitions in and out of the base states. These transitions are reversible, represented mathematically by the complex conjugate of the original transition. The set of possible state transitions is described in a matrix known as the **Hamiltonian.** A state of the system can be formed by any superposition of the base states. When used to represent binary values, one of these states is known as a **qubit**. Even a relatively small quantum system can describe a number of qubits.

Benioff described a quantum mechanical model of reversible Turing machines [Ben82]. His machine was based on a finite lattice of spin-1/2 systems. He described the state of the Turing machine in terms of a set of quantum states. He showed that each Turing machine state could be represented by a physical state as described by the system Hamiltonian. The state of the system does not degrade as the machine executes, nor does energy dissipate. The computation speed Δ—the time required for one cycle of the Turing machine—is related to the energy uncertainty. A longer Δ results in less uncertainty for the system energy. But the speed of the quantum Turing machine can be increased by increasing the average system energy; increasing system energy does not result in state degradation or energy dissipation. The efficiency of the system is given by the energy uncertainty δE divided by the computation speed Δ. That efficiency is close to the quantum limit:

$$\frac{\delta E}{\Delta} \leq 2\pi\hbar \qquad (7.3)$$

Feynman described how to build a quantum mechanical reversible computer [Fey85]. He used the reversible operations defined by Fredkin and Toffoli [Tof81;

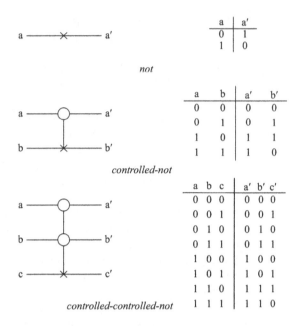

a	a'
0	1
1	0

not

a	b	a'	b'
0	0	0	0
0	1	0	1
1	0	1	1
1	1	1	0

controlled-not

a	b	c	a'	b'	c'
0	0	0	0	0	0
0	0	1	0	0	1
0	1	0	0	1	0
0	1	1	0	1	1
1	0	0	1	0	0
1	0	1	1	0	1
1	1	0	1	1	1
1	1	1	1	1	0

controlled-controlled-not

FIGURE 7.4

Two reversible primitive operators.

Fred82] and shown in Fig. 7.4: not, controlled not, and controlled-controlled-not. (This notation does not use the prime as a negation operator.) He showed how to use added atoms as program counter sites to control the sequencing of operations as governed by the Hamiltonian.

The first operational quantum computer was reported in 1998 by Jones and Mosca [Jon98]. They used nuclear magnetic resonance to execute Deutsch's algorithm, which determines whether a given function falls into one of two types, constant or balanced. Since then, several more quantum computing machines have been built of increasing complexity.

Josephson junctions   Another approach to building quantum computers relies on the **Josephson effect**. When two blocks of the proper type of material are separated by a thin conductor, a current will flow between them thanks to tunneling. Josephson received the Nobel Prize in Physics in 1973 for his prediction of this effect [Nob73]. The tunneling effect can be broken by a magnetic field. The SQUID (superconducting quantum interference device) takes advantage of the Josephson effect. A pair of Josephson junctions are arranged in parallel to create a loop in a superconducting wire. With no magnetic field, a current through the wire splits evenly through the two branches of the loop. An applied magnetic field causes a current to flow through the loop to cancel the magnetic field. This induced current reinforces the applied current on one side of the loop and cancels it on the other. As the external magnetic flux is increased, the induced current in the SQUID reverses. The induced current continues to reverse as the applied magnetic flux is increased with reversals coming at multiples of half the magnetic flux quantum. The DWave Quantum Computer [DWa16] uses SQUIDs computing elements and superconducting loops as couplers between the SQUIDs.

Entanglement

Entanglement refers to the correlations between parts of a system [Pre13]. Two photons that are entangled have correlated states—the results of measurements on the state of one photon are related to the results of measuring the state of its entangled cousin. This correlation occurs even if the photons are no longer colocated. Einstein referred to entanglement as "spooky action at a distance," but the effect has been experimentally verified using numerous experiments. However, a photon's ability to entangle with other photons is limited. The more entangled a photon becomes with another photon, the less entangled it can be with other photons.

Quantum cryptography

Quantum entanglement is useful for quantum cryptography because it can be used to detect eavesdropping. If an eavesdropper tries to read the value of a photon as it passes by, that photon's entanglement will be disturbed, causing the sender's and receiver's bits to differ. With the proper protocol, this difference can be detected and serve as a warning of eavesdroppers.

Bennett and Brassard proposed a scheme for quantum distribution of cryptographic keys [Ben84]. This scheme is designed to be resistant to eavesdropping. The sender Alice sends a photon, which can be polarized in any of four different bases. The receiver Bob receives the photon and chooses a basis in which to measure it, without knowing which basis Alice used to send it. If Bob guessed the basis in which to measure correctly, the bit will be correctly interpreted. If not, the result will be useless. If a third party tries to eavesdrop, they have a good chance of disturbing its value as received by Bob. After transmitting the bits, Alice and Bob use an ordinary communications channel to exchange information on the basis used for each bit, which allows Bob to determine which bits were correctly received. (A secure communications channel is used for this step.) Alice and Bob can then test for eavesdropping by publicly comparing some bits that should have been correctly received; those bits will no longer be secret, meaning that they cannot be used for the key. However, if Alice and Bob agree on all the bits, their communication of all the bits was likely to be secure from eavesdropping—the chance of a disagreement goes up as the number of eavesdropped bits increases. Bennett and Smolin gave the first experimental demonstration of secure quantum communication in 1989 [Smo04]. They built an apparatus that allowed pulses to be transmitted using optical fiber. The apparatus was operated in darkness to avoid contamination of the communication with stray photons.

The practical utility of quantum key exchange depends on the robustness of quantum entanglement. Entangled photons must be able to be transmitted over large distances under less-than-ideal conditions and survive for useful amounts of time. Recent experiments have shown progress on these fronts. Herbst et al. [Her15] demonstrated the teleportation of an entangled state over a distance of 143 km Tenerife to La Palma. Their quantum repeater used laser pulses transmitted through the atmosphere to perform entanglement swapping, in which two previously independent qubits are made to be entangled. Krenn et al. [Kre15] demonstrated the distribution of quantum entanglement encoded as orbital angular momentum. This property can carry a large alphabet since the amount of orbital angular momentum carried by a photon is unbounded. They demonstrated the transmission of this information in the atmosphere over a distance of 3 km using a link that allowed up to 11 quantum channels of orbital angular momentum.

7.4 Synthesis

- Carbon nanotubes can be used to build both transistors and interconnect.
- The arrangement of carbon nanotubes is relatively difficult to control, requiring self-organizing assembly.
- A Turing machine can be designed to be reversible.
- Quantum mechanical computers can implement very large state spaces in a small amount of hardware.
- Quantum entanglement can be used to detect eavesdropping of secure communications.

Useful Constants and Formulas

A.1 Physical constants

Constant	Symbol	Value
Boltzmann's constant	k	1.38×10^{-23} J/K
Charge of an electron	q	1.6×10^{-19} C
Thermal voltage at 300K	kT/q	0.026 V
Permittivity of free space	ε_0	8.854×10^{-14} F/cm
Permittivity of silicon	ε_{Si}	$11.68\varepsilon_0 = 1.03 \times 10^{-12}$ F/cm
Permittivity of SiO$_2$	ε_{ox}	$3.9\varepsilon_0 = 3.45 \times 10^{-13}$ F/cm
Concentration of carriers in intrinsic silicon	n_i	1.45×10^{10} C/cm^3
Silicon effective density of states	N_c, N_v	$N_c = 3.2 \times 10^{19}$ cm^{-3} $N_v = 1.8 \times 10^{19}$ cm^{-3}
Silicon band gap at 300K	E_g	1.12 eV

A.2 Formulas

Resistivity: $\rho = \dfrac{1}{\mu q^2 n_i}$

Resistance: $R = \rho \dfrac{l}{A}$

Carrier concentrations in doped materials:

$$n = N_d e^{-(E_d - E_f)/kT} = \frac{n_i^2}{N_a},$$

$$p = N_a e^{-(E_f - E_a)/kT} = \frac{n_i^2}{N_d}$$

Difference between Fermi levels in doped and intrinsic material: $\psi_B = \dfrac{kT}{q} \ln \dfrac{N_a}{n_i}$

Hole–electron product in equilibrium: $np = n_i^2 = N_c N_v e^{-E_g/kT}$

Shockley diode characteristic: $J = J_0\left(e^{qV/KT} - 1\right), J_0 = \dfrac{qD_n n_{p0}}{L_n} + \dfrac{qD_p p_{n0}}{L_p}$

MOS capacitor: $C_{ox} = \dfrac{\varepsilon_{ox}}{t_{ox}}, \psi_{s,inv} = 2\psi_B = 2\dfrac{kT}{q}\ln\dfrac{N_a}{n_i}$

MOSFET long-channel characteristics:

Cutoff $V_{gs} < V_t$	$I_{dn} = 0$
Linear $V_{ds} < V_{gs} - V_t$	$I_{dn} = k'_n \dfrac{W}{L}\left[(V_{gs} - V_{tn})V_{ds} - \dfrac{1}{2}V_{ds}^2\right]$
Saturation $V_{ds} \ge V_{gs} - V_t$	$I_{dn} = \dfrac{1}{2}k'_n \dfrac{W}{L}(V_{gs} - V_{tn})^2$

Subthreshold swing: $S = 2.3\dfrac{kT}{q}\left(1 + \dfrac{C_{dm}}{C_{ox}}\right)$

Rayleigh's criterion: $\mathfrak{R} = k_1 \dfrac{\lambda}{NA}$

Yield: $Y = e^{-AD}$

Middle voltage: $V_M = \dfrac{\sqrt{\dfrac{\beta_p}{\beta_n}}(V_{DD} - |V_{tp}|) + V_{tn}}{1 + \sqrt{\dfrac{\beta_p}{\beta_n}}}$

Effective resistance. $R_t = \dfrac{R_{lin} + R_{sat}}{2} = \dfrac{10V_B + 3V_t}{6\beta V_B^2}$

Delay (0–50%): $t_d = 0.69 R_t C_L$

Transition time (10–90%): $t_{rf} = 2.2 R_t C_L$

Optimal tapered driver chain: $\alpha = e, n = \ln\dfrac{C_{big}}{C_1}$

Switching energy: $E_s = C_L V_{DD}^2$

Switching power: $P_s = fC_L V_{DD}^2$

Ideal scaling: $\dfrac{\hat{t}}{t} = \dfrac{1}{x}, \dfrac{\hat{R}}{R} = x$

Thermodynamic noise error probability: $P_{err} = e^{-E_b/kT}$

Decoupling capacitance: $C_D = \dfrac{nI_{max}t_{max}}{\Delta V}$

Elmore delay for arbitrary section sizes: $\delta_E = \sum\limits_{1 \le i \le n} c_i \sum\limits_{1 \le j \le i} r_j$

Elmore delay for uniform wire: $\delta_E = \dfrac{1}{2} rcn(n+1)$

Crosstalk: $\Delta V_V = \dfrac{C_C}{C + C_C} \Delta V_A$

Clock period: $T \ge \Delta + t_s + t_h$

Metastability: $P_F = \dfrac{t_{SH}}{T} e^{-S/\tau}$

Buffered wire delay:

$$t_{bwire} = N\left[(R_b + r)c + \frac{1}{2}\left(\frac{n}{N} - 1\right)\left(\frac{n}{N} - 2\right)rc + \left(\frac{n}{N}r + R_b\right)(C_b + c)\right]$$

Bus delay: $\delta_{bus} = k_1 (C_L N)^{1/k} + k_2 N^2$

Amdahl's Law: $S(n) = \dfrac{1}{(1 - P) + \dfrac{P}{N}}$

DRAM voltages: $\dfrac{V_{bl}}{V_{line}} = \dfrac{C_{line}}{C_{line} + C_{bit}}$

Cache: $t_{av} = t_{hit} p_{hit} + t_{miss}(1 - p_{hit}) = t_{hit}[p_{hit} + M(1 - p_{hit})]$

Magnetic disk: $T_{access} = T_{seek} + T_{rot} + T_{RW}$

Paging performance: $t_{page} = t_{res}[p_{res} + M_d(1 - p_{res})]$

Relative paging performance: $\dfrac{t_{page,SSD}}{t_{page,mag}} = \dfrac{p_{res} + M_{ssd}(1 - p_{res})}{p_{res} + M_{mag}(1 - p_{res})}$

Peukart's Law: $I^n = C$

DVFS: $E_{DVFS} = nC\dfrac{G^2}{T^2}$

RTD: $E_{RTD} = n[CV^2 + L]$

Heat sink: $T_J = T_A + P\Theta$

Fourier's Law of Heat Conduction: $T = PR$

Newton's Law of Cooling: $\dfrac{dQ}{dt} = hA\Delta T,\ T(t) = T_A + (T(0) - T_A)e^{-t/t_0}$

Steady-state junction temperature: $T_J = T_A + P\Theta$

RC temperature model: $T(t) = (T_0 - PR)e^{-t/RC} + PR + T_A$

Peak-to-peak ratio: $\dfrac{T_p}{H} = \dfrac{1 - e^{-K}}{1 + e^{-K}}$

Arrhenius's equation: $r = Ae^{-E_a/kT}$

Chip life versus temperature: $\varphi_{th} = \int_0^t \dfrac{1}{kT(t)} e^{-E_a/kT(t)}$

Circuits

B.1 Introduction

This appendix is a quick review of some useful facts and methods for circuit analysis.

B.2 RLC device laws

Resistors, capacitors, and inductors are the classic electrical devices. Their schematic symbols are shown in Fig. B.1. For each of these devices, we are interested in the current I that flows through the device and the voltage V across the device's terminals.

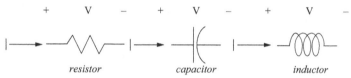

FIGURE B.1

Currents and voltages in resistors, capacitors, and inductors.

A particular resistor is characterized by its resistance. The term *resistivity* refers to the property of materials. Given a material with a given resistivity, we can build resistors of different resistance values by forming the material into different shapes.

The relationship between the voltage across a resistor and the current through it is given by **Ohm's Law**:

$$R = \frac{V}{I} \tag{B.1}$$

R is the resistance of the device.

The relationship between a capacitor's voltage and current is slightly more complex. The capacitance C is defined as

$$C = \frac{Q}{V} \tag{B.2}$$

In this case, Q is the charge stored on the capacitor. The current through the capacitor depends on the rate of change of its voltage with respect to time:

$$I = C\frac{dV}{dt} \tag{B.3}$$

233

Inductance is defined relative to the magnetic flux Φ that cuts through the inductor. The electromagnetic force on an inductor is equal to the negative of the rate of change of magnetic flux:

$$\varepsilon = \frac{d\Phi_B}{dt} \tag{B.4}$$

Inductance is defined as the ratio of flux to current:

$$L = \frac{\Phi}{I} \tag{B.5}$$

An inductor's voltage is determined by the rate of change of current as a function of time:

$$V = L\frac{dI}{dt} \tag{B.6}$$

Impedance is a generalized form of resistance that covers resistance, capacitance, and inductance. The traditional symbol for impedance is Z. An impedance can be represented as a complex number.

B.3 Circuit models

An electrical circuit is modeled as a graph. As shown in Fig. B.2, devices are represented as edges in the graph while electrical connections between the devices are represented as nodes.

The term *circuit* comes from the fact that electrical conduction requires continuous paths, which in the terminology of graphs are known as *circuits*.

B.4 Kirchhoff's laws

Kirchhoff's Current Law (KCL) and **Kirchhoff's Voltage Law (KVL)** govern the relationships between currents and voltages in circuits. KCL and KVL are very general because they are derived from conservation of mass and energy.

KCL is illustrated in Fig. B.3. In this example, three impedances form a T connection. The currents that flow through the three impedances are I_1, I_2, and I_3. In this case, we have defined the currents to all flow toward the junction; if we choose to represent

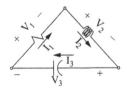

FIGURE B.2

An electrical circuit.

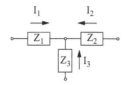

FIGURE B.3

Kirchhoff's Current Law.

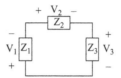

FIGURE B.4

Kirchhoff's Voltage Law.

a current as flowing in the opposite direction, we simply negate it. KCL requires that all the currents into a node must sum to zero:

$$\sum_{i \in node} I_i = 0 \qquad (B.7)$$

So in this case, $I_1 + I_2 + I_3 = 0$.

KVL is illustrated in Fig. B.4. The three impedances are arranged in a loop. Their voltages are V_1, V_2, and V_3. As with currents, we can reverse the sense of a voltage by negating it. KVL requires that the voltages around a loop to sum to zero:

$$\sum_{i \in loop} V_i = 0 \qquad (B.8)$$

So in our example, $V_1 + V_2 + V_3 = 0$.

B.5 Basic circuit analysis

We often need to find one voltage or current in the circuit given values for other voltages and currents. Due to the nature of the RLC device laws and KCL/KVL, we can use linear algebra to determine missing values in the circuit.

The voltage–current relationships in any circuit can be written in the generalized, impedance-oriented form of Ohm's Law:

$$V = ZI \qquad (B.9)$$

where

$$V = \begin{bmatrix} V_1 \\ \dots \\ V_n \end{bmatrix}, \ I = \begin{bmatrix} I_1 \\ \dots \\ I_n \end{bmatrix} \tag{B.10}$$

and

$$Z = \begin{bmatrix} Z_{11} & & Z_{1n} \\ & \ddots & \\ Z_{n1} & & Z_{nn} \end{bmatrix}. \tag{B.11}$$

The entries of the Z matrix come from the circuit graph: given two nodes i, j the value of Z_{ij} is given by the impedance between those two nodes.

Some special cases are useful for common types of circuits.

B.5.1 Series and parallel networks

We often want to represent a part of a circuit by a single **equivalent impedance**. An equivalent impedance is a form of algebraic substitution.

Fig. B.5 shows series and parallel impedances. The equivalent impedance of the series connection is

$$Z_{ser} = Z_1 + Z_2. \tag{B.12}$$

The equivalent impedance of the parallel connection is

$$Z_{\parallel} = \frac{1}{\frac{1}{Z_1} + \frac{1}{Z_2}} = \frac{Z_1 Z_2}{Z_1 + Z_2}. \tag{B.13}$$

B.5.2 Voltage dividers

A voltage divider, shown in Fig. B.6, is a very common circuit; an inverter is an example of a voltage divider. Two impedances are connected in series; given the voltage across the series connection, we want to find the voltage across one of the impedances:

$$V_b = \frac{Z_2}{Z_2 + Z_1} V_a. \tag{B.14}$$

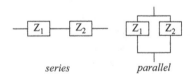

series   *parallel*

FIGURE B.5

Series and parallel impedances.

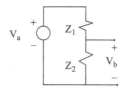

FIGURE B.6

A voltage divider.

B.5.3 **Ladder networks**

Ladder networks are a common form and are amenable to algebraic analysis by hand. The transmission line is an example of a ladder network.

Fig. B.7 shows a two-section ladder circuit. Its input is at V_0 and its output is at V_4. We can find V_4 by transforming the latter part of the network into a single equivalent impedance.

$$Z_{234} = Z_2 \parallel (Z_3 + Z_4) = \frac{Z_2(Z_3 + Z_4)}{Z_2 + Z_3 + Z_4}. \tag{B.15}$$

We can now easily find V_2 using the voltage divider:

$$V_2 = V_0 \frac{Z_{234}}{Z_1 + Z_{234}}. \tag{B.16}$$

Given V_2, we can use the voltage divider relation again to find V_4:

$$
\begin{aligned}
V_4 &= V_2 \frac{Z_2}{Z_3 + Z_4} \\
&= V_0 \frac{Z_{234}}{Z_1 + Z_{234}} \frac{Z_2}{Z_3 + Z_4} \\
&= V_0 \frac{\frac{Z_2(Z_3 + Z_4)}{Z_2 + Z_3 + Z_4}}{Z_1 + \frac{Z_2(Z_3 + Z_4)}{Z_2 + Z_3 + Z_4}} \frac{Z_2}{Z_3 + Z_4}
\end{aligned}
\tag{B.17}
$$

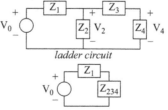

ladder circuit

equivalent reduced circuit

FIGURE B.7

Analysis of a two-section ladder network.

B.6 Differential equations and circuits

To fully understand the behavior of circuits with capacitors and inductors, we need to introduce time into our circuit equations. As we saw above, the laws that govern capacitors and inductors are differential equations. When we combine several impedances into a circuit, we can use KCL and KVL to help us write the differential equations that govern the circuit.

Fig. B.8 shows a simple RC circuit with a resistor and capacitor connected in series. We need an initial condition to determine its behavior over time; we will assume that the charge on the capacitor at time $t = 0$ is Q_0.

KVL tells us that the sum of the voltages around the circuit is zero:

$$V_R + V_C = 0 \tag{B.18}$$

We can substitute the resistor and capacitor laws into this equation:

$$\frac{dq}{dt}R + \frac{q(t)}{C} = 0. \tag{B.19}$$

Because the capacitor law is written in terms of charge q, we have used the fact that current is the flow of charge per unit time to rewrite the current through the resistor as $I = \dfrac{dq}{dt}$.

We can integrate this equation to find an expression for the capacitor's charge as a function of time:

$$\frac{dq}{dt}R + \frac{q(t)}{C} = \int_0^q \frac{1}{q}\,dq + \frac{1}{RC}\int_0^t dt = \ln\frac{q}{Q_0} + \frac{t}{RC} \tag{B.20}$$

$$q(t) = Q_0 e^{-t/RC} \tag{B.21}$$

To find the current as a function of time, we differentiate the charge with respect to time:

$$I(t) = -\frac{Q_0}{RC}e^{-t/RC} = -\frac{V_0}{R}e^{-t/RC} \tag{B.22}$$

Given this expression for the current, we can easily find the voltage across the resistor, which then gives us the capacitor voltage using KVL:

$$V_C(t) = V_0 e^{-t/RC} \tag{B.23}$$

FIGURE B.8

An RC circuit.

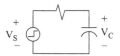

FIGURE B.9

An *RC* circuit with a voltage step source.

We generally use the symbol $\tau = RC$ which is known as the **time constant**. This quantity has units of time and gives us a quick reference for the speed of the circuit: when $t = \tau$, then $e^{-t/RC} = e^{-\tau/\tau} = 0.37$.

Fig. B.9 shows a circuit with a voltage step input source. The voltage source changes from 0 to V_s at $t = 0$. The initial voltage of the capacitor is $V_c(0) = V_0$. The form of the capacitor voltage waveform for $t \geq 0$ is

$$V_c(t) = (V_0 - V_s)e^{-t/RC} + V_s. \tag{B.24}$$

B.7 Linear time-invariant systems

An *RC* circuit is an example of a **linear time-invariant (LTI) system**. LTI systems are amenable to hand analysis.

A system is linear if it obeys **superposition**:

$$f(a + b) = f(a) + f(b) \tag{B.25}$$

Superposition allows us to find the circuit's response to a complex signal by breaking the signal into simpler signals and finding the circuit's response to those signals. Intuitively, a linear system's input/output graph is a slope through the origin as shown in Fig. B.10.

Transistors and diodes are nonlinear. We may be able to build linear models of these devices that hold in a particular region of operation, but their overall behavior is nonlinear. Solving the circuit equations involving nonlinear devices requires numerical methods.

A time-invariant system is one whose parameters—for example, resistance—do not change. Time invariance is a useful property of ideal components. However, real components vary with temperature, aging.

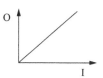

FIGURE B.10

Form of a linear system.

Questions

QB-1 Solve for the currents in this circuit.

QB-2 Solve for the voltages in this circuit.

QB-3 If V_1 is 1.0 V, what is V_2?

QB-4 Derive Eq. (B.14) for the voltage divider.

QB-5 Derive Eq. (B.15) for the ladder stage equivalent impedance.

QB-6 You are given an RC circuit with a step voltage source of 1 V connected in series with a resistor R and capacitor C. The initial voltage of the capacitor is 1 V. Find an expression for the capacitor voltage V_c as a function of time.

QB-7 An RC circuit is given a step input with the output voltage measured across the capacitor. Find the time required for the output voltage to rise from 0% to 50% of its final value.

QB-8 6 Use superposition to find V_2 in the circuit of QB-2 if the voltage source value is 2 V.

Probability

C.1 Introduction

An intuitive formulation of probability is based on a set of discrete events, some of which are favorable and others unfavorable. The probability of a favorable event is

$$P(F) = \text{Pr}\{favorable\} = \frac{n_{favorable}}{n_{favorable} + n_{unfavorable}} \tag{C.1}$$

This definition implies $0 \leq P \leq 1$. Odds, by comparison, are the ratio of favorable to unfavorable.

Two events are **independent** if the outcome of one does not affect the outcome of another. The joint probability of two independent events is

$$P(A \text{ and } B) = P(A)P(B). \tag{C.2}$$

Two events are **mutually exclusive** if they cannot both happen. In this case,

$$P(A \text{ xor } B) = P(A) + P(B). \tag{C.3}$$

If two events are not mutually exclusive, then

$$P(A \text{ or } B) = P(A) + P(B) - P(A \text{ and } B). \tag{C.4}$$

A **conditional probability** is the probability of an event A given that event B occurred:

$$P(A|B) = \frac{P(A \text{ and } B)}{P(B)} \tag{C.5}$$

The **probability mass function (PMF)** is the probability that a discrete random variable takes on a particular value. The **probability density function (PDF)** is the likelihood that a continuous random variable takes on a given value. The probability of a value in given range is given by the integral of the PDF, known as the **cumulative distribution function (CDF)**.

C.2 **Poisson distribution**

The **Poisson distribution** describes the probability of observing k events in a given interval:

$$P(N = k) = \frac{\lambda^k e^{-\lambda}}{k!}. \tag{C.6}$$

λ is the **event rate**, the average number of events in an interval. The Poisson distribution assumes that events are independent. A useful special case is for no events to happen:

$$P(N = 0) = e^{-\lambda}. \tag{C.7}$$

C.3 **Exponential distribution**

The exponential distribution describes the time between events in a Poisson process. The exponential PDF is

$$P(x) = \lambda e^{-\lambda x}, x \geq 0. \tag{C.8}$$

In this formula, λ is the Poisson process's event rate. The CDF gives the probability of an event occurring in the interval $[0, x]$:

$$F(x) = 1 - e^{-\lambda x}, x \geq 0. \tag{C.9}$$

C.4 **Gaussian distribution**

The **Gaussian distribution** is also known as the normal distribution. Given a **mean** μ and **variance** σ for events, the Gaussian PDF is

$$P(x) = \frac{1}{\sigma\sqrt{2\pi}} e^{-\frac{(x-\mu)^2}{2\sigma^2}}. \tag{C.10}$$

Advanced Topics

D.1 Introduction

In this appendix we briefly discuss several advanced topics in CMOS digital design. Section D.2 describes some advanced device characteristics, Section D.3 concentrates on gate delay, and Section D.4 looks at interconnect delay.

D.2 Device characteristics

p-n junctions

The width of the depletion layer of a diode is [Tau98]

$$W_d = \sqrt{\frac{2\varepsilon_{si}(N_a + N_d)\psi_m}{qN_aN_d}} \tag{D.1}$$

where ψ_m is the potential drop across the *p-n* junction. The capacitance of the depletion layer is

$$C_d = \frac{\varepsilon_{si}}{W_d}. \tag{D.2}$$

MOSFET

Body effect is the relationship between substrate bias V_{bs} and threshold voltage. The substrate sensitivity is [Tau98]

$$\frac{dV_t}{dV_{bs}} = \frac{\sqrt{\frac{\varepsilon_{si}qN_a2}{(2\psi_b + V_{bs})}}}{C_{ox}}. \tag{D.3}$$

Short-channel effects

Short-channel devices require more complex models [Tau98]. In these cases, the distance between the source and drain is comparable to the width of the source/drain junction depletion regions. As a result, the source and drain regions affect the band structure over much of the channel.

Drain-induced barrier lowering is one result of short-channel phenomena. The incursion of the depletion regions into the channel lowers the threshold voltage and reduces the potential barrier, increasing leakage current. High drain voltages further lower the barrier, resulting in larger decreases in the threshold voltage and increases in leakage current.

Drain current saturates at a lower voltage due to velocity saturation of the carriers. The saturation drain current is also modulated. The distance between the saturation pinchoff point and the drain boundary is ΔL; the saturation drain current becomes

$$I_d = \frac{I_{d,sat}}{1 - (\Delta L/L)}. \tag{D.4}$$

Increasing drain voltage increases ΔL, resulting in increased drain currents.

243

D.3 Logic gates

We will first introduce topologies for noninverter CMOS gates. We will then describe two models for gate delay: the Horowitz model in Section D.3.1 and the Sakurai–Newton model in Section D.3.2.

D.3.1 Gate topologies

Static complementary logic is static because it does not rely on charge storage; it is complementary because the pullup and pulldown networks implement complementary switching functions.

Fig. D.1 shows the topology of two-input NAND and NOR gates. The terms AOI (AND-OR-INVERT) and OAI (OR-AND-INVERT) are used for complex gates.

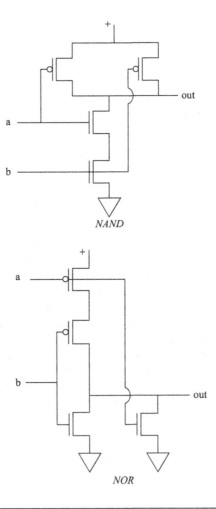

FIGURE D.1

Two-input NAND and NOR complementary logic gates.

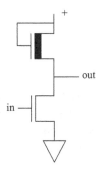

FIGURE D.2

An nMOS inverter.

Ratioed logic

Static CMOS is ratioless because its operation does not depend on a ratio of device parameters. Ratioless operation allows CMOS to operate over a wide range of power supply voltages. Many logic families are ratioed. The nMOS gate shown in Fig. D.2 is an example. The pullup transistor is a depletion mode transistor that is always on and can be modeled as a resistor. On a rising output, the pullup transistor pulls the output voltage to the power supply. However, on a falling transition, both the pullup and pulldown remain on. The steady-state output voltage is determined by the voltage divider formed by the pullup and pulldown.

A dynamic logic gate makes use of stored charge. Fig. D.3 shows the topology of a domino logic gate. A clock signal controls evaluation of the gate, which operates over two phases. When the clock is low, the pullup transistor precharges the inverter input node. When the clock is high, the pullup switches off and the clocked pulldown transistors allow the gate to be evaluated. The term *domino* comes from the requirement

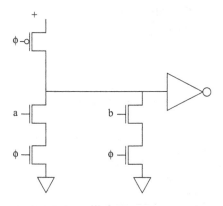

FIGURE D.3

A domino logic gate.

that all logic signals in domino logic must monotonically increase. A glitch on a logic signal when the clock is high will cause the precharge node to be discharged. Since that node cannot be recharged during the evaluation period, the gate's output is corrupted.

D.3.2 Horowitz slope-dependent delay model

Horowitz [Hor84] developed a model for input-slope-dependent gate delay. We can model slope-dependent delay by using a current source for a high-gain region and a resistor for a low-gain region. The gate has a switching voltage V_s. Its transcondutance in the saturation region is g_m; in this region the output current is $V_{in}g_m$. The gate's output resistance in the low-gain region is R_f. If we model the input waveform as a rising ramp (in the case of a $0 \rightarrow 1$ input), we can write its form as

$$V_{in}(t) = V_s + \frac{t}{\alpha \tau_f} \tag{D.5}$$

where $\tau_f = R_f C_L$. At the start of the transition, the gate operates in the high-gain region. The output voltage in this region is

$$V_{out}(t) = \frac{t^2}{2\alpha\beta\tau_f^2} \tag{D.6}$$

where $\beta = 1/R_f C_L$. The gate enters the low-gain region at

$$t_s = \tau_f \left[\sqrt{1 + 2\alpha\beta} - 1 \right]. \tag{D.7}$$

The output voltage in this region is

$$V_{out}(t) = \left(1 - \frac{t^2}{2\alpha\beta\tau_f^2} \right) e^{(t_s - t)/\tau_f} \tag{D.8}$$

We will define t_d to be the time from when the input reaches the V_s to when the output voltage reaches V_s. This value can be written as

$$t_d = \sqrt{(\tau_f \ln V_s)^2 + 2\tau_{in}\tau_g(1 - V_s)} \tag{D.9}$$

The first term inside the square root is the intrinsic gate delay, and the second term is the slope-dependent delay.

D.3.3 Sakurai–Newton model

Power law model

The **Sakurai–Newton power law model** [Sak91; Sak91B] is an important tool for circuit analysis. It is an analytical model for delay that can be used to derive expressions for delay that allow us to understand the relationships between parameters. It captures a number of transistor phenomena as well as realistic input waveforms.

The transistor model is a generalization of our long-channel MOSFET model that has been modified to take into account several important short-channel effects. It relies on several parameters, most of which are measured rather than derived from first principles. First, some transistor parameters [Sak91B]:

$$V_t = V_{t0} + \gamma \left(\sqrt{2\varphi_f - V_{bs}} - \sqrt{2\varphi_f} \right), \tag{D.10}$$

$$V_{d\,sat} = K(V_{gs} - V_t)^m, \tag{D.11}$$

$$I_{d\,sat} = \frac{W}{L_{eff}} B(V_{gs} - V_t)^n. \tag{D.12}$$

V_{gs}, V_{ds}, and V_{bs} are the gate-source, drain-source, and substrate-source (or bulk-source) voltages. W is the device width. L_{eff} is the effective channel length taking into account effects that cause the channel to behave as if it is shorter than its geometric length. V_t is the threshold voltage and $V_{d\,sat}$ the drain saturation voltage, while V_{t0}, γ and $2\varphi_f$ are parameters of the threshold voltage. $I_{d\,sat}$ is the drain saturation current. K and m describe the linear region and B and n describe the saturation region. λ_0 and λ_1 describe the saturation region's finite conductance.

The drain current equations model linear (I_{d3}) and saturation (I_{d5}) regions as with the long-channel model:

$$I_{d3} = I_{d5} \left(2 - \frac{V_{ds}}{V_{d\,sat}} \right) \frac{V_{ds}}{V_{d\,sat}}, \tag{D.13}$$

$$I_{d5} = I_{d\,sat}(1 + \lambda V_{ds}). \tag{D.14}$$

where $\lambda = \lambda_0 - \lambda_1 V_{bs}$. The boundary between the linear and saturation regions is

$$V_{ds} < V_{d\,sat}. \tag{D.15}$$

In the cutoff region, the drain current is 0.

To use these equations to analyze circuit delay, the behavior is divided into two regions: fast input and slow input. The curves of these slow and fast regions are smoothly connected, simplifying our analysis. The boundary between the two is the critical input transition time [Sak91]:

$$t_{T0} = \frac{C_o V_{DD}}{2 I_{do}} \frac{(n+1)(1 - v_t)^n}{(1 - v_t)^{n+1} - (v_v - v_t)^{n+1}} \tag{D.16}$$

In this formula, $v_t = V_{t0}/V_{DD}$, $v_v = V_{inv}/V_{DD}$; C_o is the output capacitance of the gate and I_{do} is the output drain current.

Unlike our simple *RC* model, the power law model does not assume a step input. It models the input as a waveform shown in Fig. D.4: the input voltage is 0 until the input reaches the logic threshold voltage $V_{inv} = v_v V_{DD}$, then jumps up to $V_{in,ap}$, then ramps up to the power supply voltage with a ramp slope of $V_{in,ap}/V_{inv}$. The model does assume, as we have done before, that only one transistor is on at a time.

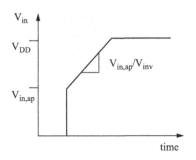

FIGURE D.4

Sakurai–Newton model for gate input voltage.

The delay model is broken into two cases. The delay time t_d measures the time between the input reaching 50% of the power supply voltage and the output reaching the same 50% level. The output transition time $t_{T\,out}$ can be used as the transition time t_T for the next gate's input. The fast input model covers the case in which the input transition time $t_T \leq t_{T0}$:

$$t_d = \tau_T \left\{ \frac{1}{2} - \frac{1 - v_t}{n+1} + \frac{(v_v - v_t)^{n+1}}{(n+1)(1 - v_t)^n} \right\} + \frac{C_o V_{DD}}{2 I_{do}}, \tag{D.17}$$

$$t_{T\,out} = \frac{C_o V_{DD}}{0.7 I_{do}} \frac{4 v_{d0}^2}{4 v_{d0} - 1}. \tag{D.18}$$

For the slow input case $t_T > t_{T0}$:

$$t_d = t_T \left[v_t - \frac{1}{2} + \left\{ (v_v - v_t)^{n+1} + \frac{(n+1)(1 - v_t)^n}{2 t_T I_{do}/C_o V_{DD}} \right\}^{1/n+1} \right], \tag{D.19}$$

$$t_{T\,out} = \frac{C_o V_{DD}}{0.7 I_{do}} \left(\frac{1 - v_t}{t_d/t_T + 1/(2 - v_t)} \right)^n. \tag{D.20}$$

$v_{d0} = V_{d0}/V_{DD}$ in these formulas.

Power law applications

Sakurai and Newton were able to show some interesting properties of logic gate circuits using this model [Sak91], particularly for gates with several series transistors as are found in NANDs, NORs, and complex gates. The *RC* model predicts that when the load capacitance is small, delay should be a quadratic function in the number of transistors *N*; they showed that delay increases at a factor between linear and quadratic. When the load capacitance is large, the *RC* model predicts delay to be linear in *N*; they verified that this is the case even for their more complex short-channel model. They also considered the problem of which input to a logic gate is the slowest—the one closest to the output or farthest away? This problem is particularly important in adder design. They showed that for small load capacitance, the input closest to the gate's output is fastest, while for large load capacitance, the input farthest from the gate is fastest.

D.4 Interconnect

In this section we discuss the analysis of interconnect: *RC* tree delay in Section D.4.1,
buffer placement in *RC* wires in Section D.4.2, and RLC tree delay in Section D.4.3.

D.4.1 *RC* trees

RC trees are a generalization of the *RC* transmission line in which a gate drives several
loads. Rubenstein, Penfield, and Horowitz [Rub83] developed upper and lower bounds
on the step response of *RC* trees; their model is commonly known as the Penfield–Rubenstein model. Their analysis makes use of several sums over parts of the
RC tree:

$$T_P = \sum_k R_{kk}C_k, \tag{D.21}$$

$$T_{Di} = \sum_k R_{ki}C_k, \tag{D.22}$$

$$T_{Ri} = \sum_k R_{ki}^2 C_k \Big/ R_{ii}. \tag{D.23}$$

In these formulas, C_k is the capacitance at node k, R_{kk} is the total resistance from the
input to node k, and R_{ki} is the resistance along the intersections of the paths from
the input to node k and node i. T_P is the sum of the open-circuit time constants; T_{Di} is
the first-order moment of the impulse response. This function is used to derive the
bounds:

$$f_i(t) = \int_0^t [1 - v(t')]dt' \tag{D.24}$$

$f_i(\infty) = T_{Di}$, the first-order impulse response moment. Rubeinstein et al. showed
that

$$T_{Ri}[1 - v_i(t)] \le T_{Di} - f_i(t) \le T_P[1 - v_i(t)] \tag{D.25}$$

Their voltage bounds are shown in Table D.1, and their time bounds are shown in
Table D.2.

Table D.1 RPH Bounds on Voltage [RPH83]

Lower	$0, t \le T_{Di} - T_{Ri}$
	$1 - \dfrac{T_{Di}}{t + T_{Ri}}, T_{Di} - T_{Ri} \le t \le T_P - T_{Ri}$
	$1 - \dfrac{T_{Di}}{T_P} e^{(T_P - T_{Ri})/T_P} e^{-t/T_P}, t \ge T_P - T_{Ri}$
Upper	$1 - \dfrac{T_{Di} - t}{T_P}, t \le T_{Di} - T_{Ri}$
	$1 - \dfrac{T_{Ri}}{T_P} e^{(T_{Di} - T_{Ri})/T_{Ri}} e^{-t/T_{Ri}}, t \ge T_{Di} - T_{Ri}$

Table D.2 Penfield–Rubenstein Bounds on Time [Rub83]

Lower	$0, \; v_i(t) \leq 1 - \dfrac{T_{Di}}{T_P}$
	$T_{Di} - T_P[1 - v_i(t)], \; 1 - \dfrac{T_{Di}}{T_P} \leq v_i(t) \leq 1 - \dfrac{T_{Ri}}{T_P}$
	$T_{Di} - T_{Ri} + T_{Ri} \ln \dfrac{T_{Di}}{T_P[1 - v_i(t)]}, \; v_i(t) \geq 1 - \dfrac{T_{Ri}}{T_P}$
Upper	$\dfrac{T_{Di}}{1 - v_i(t)}, \; v_i(t) \leq 1 - \dfrac{T_{Di}}{T_P}$
	$T_P - T_{Ri} + T_P \ln \dfrac{T_{Di}}{T_P[1 - v_i(t)]}, \; v_i(t) \geq 1 - \dfrac{T_{Di}}{T_P}$

D.4.2 Interconnect buffering

Bakoglu [Bak90] analyzed the placement and sizing of buffers in RC transmission lines. Given an RC transmission line with total impedance R_{int}, C_{int}, the problem is to divide the line into k sections with a buffer between each pair of sections (as well as at the two ends of the transmission line); each section is of length l.

If all buffers are of size 1, then the 50% delay is

$$\delta = k\left[0.7R_0\left(\frac{C_{int}}{k} + C_0\right) + \frac{R_{int}}{k}\left(0.4\frac{C_{int}}{k} + 0.7C_0\right)\right] \tag{D.26}$$

where R_0, C_0 are the driver's equivalent resistance and input capacitance. The optimal number of repeaters is found at $d\delta/dk = 0$:

$$k = \sqrt{\frac{0.4R_{int}C_{int}}{0.7R_0C_0}}. \tag{D.27}$$

This result can be generalized to a repeater size h. In this case,

$$\delta = k\left[0.7\frac{R_0}{h}\left(\frac{C_{int}}{k} + hC_0\right) + \frac{R_{int}}{k}\left(0.4\frac{C_{int}}{k} + 0.7hC_0\right)\right], \tag{D.28}$$

$$k = \sqrt{\frac{0.4R_{int}C_{int}}{0.7R_0C_0}}, \tag{D.29}$$

$$h = \sqrt{\frac{R_0C_{int}}{R_{int}C_0}}. \tag{D.30}$$

At these values,

$$\delta = 2.5\sqrt{R_0C_0R_{int}C_{int}}. \tag{D.31}$$

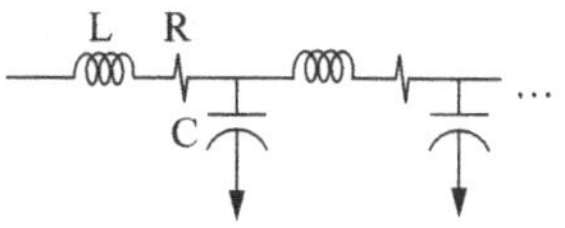

FIGURE D.5

An RLC transmission line.

D.4.3 Inductive interconnect and RLC trees

RLC basics

Inductive interconnect has much more complex properties than does strictly capacitive interconnect. Inductive effects become more important as resistance decreases and frequency increases.

Fig. D.5 shows a model of an RLC line. The poles of the RLC section are located at

$$\omega_0 \left[\xi \pm \sqrt{\xi^2 - 1} \right] \tag{D.32}$$

where

$$\omega_0 = \frac{1}{\sqrt{LC}}, \tag{D.33}$$

$$\xi = \frac{R}{2} \sqrt{\frac{C}{L}}. \tag{D.34}$$

ξ is the damping factor. If $\xi > 1$, the circuit is **overdamped** and its impulse response is the sum of two exponentials. If $\xi < 1$, the circuit is **underdamped** and its response is an exponentially tapered sinusoid.

An LC transmission line with no resistance gives a lower bound on propagation delay in RLC interconnect. The velocity signal propagation of an ideal LC transmission line is [Ram65]

$$v = \frac{1}{\sqrt{LC}}. \tag{D.35}$$

RLC trees

The Penfield–Rubenstein approach can be generalized to *RC* trees [Ism00; Sal12]. Two moments are required for each RLC node:

$$m_1^i = - \sum_k C_k R_{ki} \tag{D.36}$$

$$m_2^i = \left(\sum_k C_k R_{ki} \right)^2 - \sum_k C_k L_{ki} \tag{D.37}$$

The damping factor and natural frequency for node i can be written as

$$\xi_i = \frac{1}{2} \frac{\sum\limits_k C_k R_{ki}}{\sqrt{\sum\limits_k C_k L_{ki}}} \tag{D.38}$$

$$\omega_{ni} = \frac{1}{\sqrt{\sum\limits_k C_k L_{ki}}} \tag{D.39}$$

The 50% delay for node i is

$$t_{pdi} = \frac{1.047 e^{\xi_i/0.85}}{\omega_{ni}} + 0.695 \sum_k C_k R_{ki}. \tag{D.40}$$

References

[Ade94] L. M. Adelman, "Molecular computation of solutions to combinatorial problems," *Science*, November 11, 1994, pp. 1021–1024.

[Aga07] Vishwani D. Agrawal and Srivaths Ravi, "Low-power design and test: dynamic and static power in CMOS," July 30–31, 2007, www.eng.auburn.edu/~agrawvd/...07.../lp_hyd_2.ppt.

[All85] Ross R. Allen, John D. Meyer, and William R. Knight, "Thermodynamics and hydrodynamics of thermal ink jet printers," *Hewlett-Packard Journal*, 36(5), May 1985, pp. 21–27.

[Ata60] M. M. Atalla, "Semiconductor devices having dielectric coatings," U.S. Patent 3,206,670, March 8, 1960.

[Avo03] Phaedon Avouris, Joerg Appenzeller, Richard Martel, and Shalom J. Wind, "Carbon nanotube electronics," *Proceedings of the IEEE*, 91(11), November 2003, pp. 1772–1784.

[Bac01] Adrian Bachtold, Peter Hadley, Takeshi Nakanishi, and Cees Dekker, "Logic circuits with carbon nanotube transistors," *Science*, 294(5545), November 9, 2001, pp. 1317–1320, http://dx.doi.org/10.1126/science.1065824.

[Bak90] H. B. Bakoglu, *Circuits, Interconnections, and Packaging for VLSI*, Boston: Addison-Wesley, 1990.

[Bar50] John Bardeen and Walter H. Brattain, "Three-electrode circuit element utilizing semiconductor materials," U.S. Patent 2,524,035, October 3, 1950.

[Bay75] Bryce E. Bayer, "Color imaging array," U.S. Patent 3,971,065, March 5, 1975.

[Ben73] Charles H. Bennett, "Logical reversibility of computation," *IBM Journal of Research and Development*, 17(6), November 1973, pp. 525–532.

[Ben82] Paul Benioff, "Quantum mechanical models of Turing machines that dissipate no energy," *Physical Review Letters*, 48(23), June 7, 1982, pp. 1581–1585.

[Ben84] C. H. Bennett and G. Brassard, "Quantum cryptography: public-key distribution and coin tossing," *Proceedings of the IEEE International Conference on Computers, Systems, and Signal Processing, Bangalore, India*, 1984, pp. 175–179.

[Bla69] J. R. Black, "Electromigration—a brief survey and some recent results," In: *IEEE Transactions on Electron Devices*, 16(4), April 1969, pp. 338–347.

[Bud04] Ravi Budruk, Don Anderson, and Tom Shanley, *PCI Express System Architecture*, Boston: Addison-Wesley, 2004.

[Cha73] Thomas J. Chaney and Charles E. Molnar, "Anomalous behavior of synchronizer and arbiter circuits," *IEEE Transactions on Computers*, C-22(4), April 1973, pp. 421–422.

[Coe16] David Coelho, personal communication, January 29, 2016.

[Cyp15] Cypress, *PSoC 5LP: CY8C52LP Family Datasheet*, Document number 001-84933, Rev. I, November 30, 2015.

[Deh03] Andre DeHon, "Array-based architecture for FET-based, nanoscale electronics," *IEEE Transactions on Nanotechnology*, 2(1), March 2003, pp. 23–32.

[Den68] Robert H. Dennard, "Field-effect transistor memory," U.S. Patent 3,387,286, June 4, 1968.

[Den74] Robert H. Dennard, Fritz H. Gaensslen, Hwa-Nien Yu, V. Leo Rideout, Ernest Bassous, and Andre R. LeBlanc, "Design of ion-implanted MOSFET's with very small physical dimensions," *IEEE Journal of Solid-State Circuits*, SC-9(5), October 1974, pp. 256–268.

[DWa16] DWave, "Introduction to the D-Wave Quantum Hardware," http://www.dwavesys.com/tutorials/background-reading-series/introduction-d-wave-quantum-hardware.

[Edi79] Thomas A. Edison, *Improvement in Carbon-Telephones*, U.S. Patent 222,390, December 9, 1879.

[Edi82A] Thomas A. Edison, *Telephone*, U.S. Patent 252,422, January 17, 1882.

[Edi82B] Thomas A. Edison, *Telephone*, U.S. Patent 266,022, October 17, 1882.

[Edi84] Thomas A. Edison, *Electrical indicator*, U.S Patent 307,031, October 21, 1884.

[EIA15] U.S. Energy Information Agency, "Frequently Asked Questions," http://www.eia.gov/tools/faqs/faq.cfm?id=97&t=3, October 15, 2015.

[ElG05] Abbas El Gamal and Helmy Eltoukhy, "CMOS image sensors," *IEEE Circuits and Devices Magazine*, May/June 2005, pp. 6–20.

[Elm48] W. C. Elmore, "The transient response of damped linear networks with particular regard to wideband amplifiers," *Journal of Applied Physics*, 19, 1948, 55–63.

[Fey85] Richard P. Feynman, "Quantum mechanical computers," *Optics News*, February 1985, pp. 11–20.

[Fey10A] Richard P. Feynman, Robert B. Leighton, and Matthew Sands, *The Feynman Lectures on Physics, Volume I: Mainly Mechanics, Radiation and Heat*, Millenium Edition, New York: Basic Books, 2010.

[Fey10B] Richard P. Feynman, Robert B. Leighton, and Matthew Sands, *The Feynman Lectures on Physics, Volume II: Mainly Electromagnetism and Matter*, Millenium Edition, New York: Basic Books, 2010.

[Fis95] J. P. Fishburn and C. A. Schevon, "Shaping a distributed-RC line to minimize Elmore delay," *IEEE Transactions on CAS-I*, 42, December, 1995, pp. 1020–1022.

[Fla85] Stephen T. Flannagan, "Synchronization reliability in CMOS technology," *IEEE Journal of Solid-State Circuits*, 20(4), August 1985, pp. 880–882.

[Fos95] Eric R. Fossum, "CMOS image sensors: electronic camera on a chip," In: *International Electron Devices Meeting, 1995*, IEEE, 1995, pp. 17–25.

[Fle05] J. A. Fleming, "Instrument for converting alternating currents into continuous currents," U.S. Patent 803,684, November 7, 1905.

[Fre82] Edward Fredkin and Tommaso Toffoli, "Conservative logic," *International Journal of Theoretical Physics*, 21(3/4), 1982, pp. 219–253.

[Gar08] Martin Gardner, *Origami, Eleusis, and the Soma Cube*, Cambridge University Press, 2008.

[GEL15] GE Lighting, "News – First LED by the GE engineer, Nick Holonyak," http://www.gelighting.com/LightingWeb/emea/news-and-media/news/First-LED-by-the-GE-engineer-Nick-Holonyak.jsp.

[Gin11] Ran Ginosar, "Metastability and synchronizers: a tutorial," *IEEE Design & Test of Computers*, 28(5), September/October 2011, pp. 23–35.

[Gup08] Puneet Gupta and Evanthia Papadopoulou, "Yield analysis and optimization," Chapter 37 In: Charles J. Alpert, Dinesh P. Mehta, and Sachin S. Sapatnekar (Eds.), *Handbook of Algorithms for Physical Design Automation*, Auerbach Publications, 2008.

[Hal88] David Halliday, Robert Resnick, and John Merrill, *Fundamentals of Physics, Third Edition Extended*, New York: John Wiley and Sons, 1988.

[Hec98] Eugene Hecht, *Optics*, third edition, Addison Wesley Longman, 1998.

[Hei68] George H. Heilmeier, Louis A. Zanoni, and Lucian A. Barton, "Dynamic scattering: a new electrooptic effect in certain classes of nematic liquid crystals," *Proceedings of the IEEE*, 56(7), July 1968, pp. 1162–1171.

[Her15] Thomas Herbst, Thomas Scheidl, Matthias Fink, Johannes Handsteiner, Bernhard Wittmann, Rupert Ursin, and Anton Zeilinger, "Teleportation of entanglement over 143 km," *PNAS*, 112(46), 2015, 14202−14205; published ahead of print November 2, 2015, http://dx.doi.org/10.1073/pnas.1517007112.

[HGS12] HGST, *UltraStar A7K2000 3.5-Inch Enterprise 7200 RPM Hard Disk Drives*, DSUA722009EN-02, 2012.

[Hoe62] Jean A. Hoerni, "Method of manufacturing semiconductor devices," U.S. Patent 3,025,589, March 20, 1962.

[Hol88] Herman Hollerith, "Art of compiling statistics," U.S. Patent 395,782, September 8, 1888.

[Hor84] Mark Alan Horowitz, *Timing Models for MOS Circuits*, prepared under U.S. Army Research Office Contract No. DAAG-29-80-K-0046, Integrated Circuits Laboratory, Stanford Electronics Laboratories, Stanford University, Stanford, California, January 1984.

[Int11] Intel, *Intel Xeon Processor E7-8800/4800/2800 Product Families, Datasheet Volume 1 of 2*, Reference Number 325119-001, April 2011.

[Int12] Intel, *Intel Solid-State Drive 520 Series Product Specification*, Order Number 325986-001US, February 2012.

[Int14] Intel, *PHY Interface for the PCI Express, SATA, and USB 3.1 Architectures*, version 4.3, 2014.

[Ish16] Alex Ishii, personal communication, April 7, 2016.

[Ism00] Y. I. Ismail, E. G. Friedman and J. L. Neves, "Equivalent Elmore delay for RLC trees," In: *IEEE Transactions on Computer-Aided Design of Integrated Circuits and Systems*, vol. 19, no. 1, January 2000, pp. 83−97.

[ITR01] ITRS, *International Technology Roadmap for Semiconductors: 2001 Edition, Executive Summary*, 2001, http://itrs.net.

[ITR05] ITRS, *International Technology Roadmap for Semiconductors: 2005 Edition, Executive Summary*, 2005, http://itrs.net.

[ITR07] ITRS, *International Technology Roadmap for Semiconductors: 2007 Edition, Executive Summary*, 2007, http://itrs.net.

[ITR11] ITRS, *International Technology Roadmap for Semiconductors: 2011 Edition, Executive Summary*, 2011, http://itrs.net.

[ITR13] ITRS, *International Technology Roadmap for Semiconductors: 2013 Edition, IRC Overview*, 2013, http://itrs.net.

[Jae75] Richard C. Jaeger, "Comments on 'An optimized output stage for MOS integrated circuits,'" *IEEE Journal of Solid-State Circuits*, SC10(3), June 1975, pp. 185−186.

[Jed16] JEDEC Solid State Technology Association, *High Bandwidth Memory (HBM) DRAM*, JESD235A, November 2015.

[Jon98] A. Jones and M. Mosca, "Implementation of a Quantum Algorithm on a Nuclear Magnetic Resonance Quantum Computer," *Journal of Chemical Physics*, 109, 1998, pp. 1648−1653.

[Kah63] Dawon Kahng, "Electric field controlled semiconductor device," U.S. Patent 3,102,230, August 27, 1963.

[Kah67] D. Kahng and S. M. Sze, "A floating gate and Its application to memory devices," *Bell System Technical Journal*, 46, 1967, pp. 1288−1295.

[Kah76] Dahwon Kahng, "A historical perspective on the development of MOS transistors and related devices," *IEEE Transactions on Electron Devices*, 23(7), July 1976, pp. 655−657.

[Key70] R. W. Keyes, and R. Landauer, "Minimal energy dissipation in logic," *IBM Journal of Research and Development*, vol. 14, no. 2, March 1970, pp. 152–157.

[Kil64] Jack S. Kilby, "Miniaturized electronic circuits," U.S. Patent 3,138,743, June 23, 1964.

[Kre15] Mario Krenn, Johannes Handsteiner, Matthias Fink, Robert Fickler, and Anton Zeilinger, "Twisted photon entanglement through turbulent air across Vienna," PNAS, 2015 112(46), pp. 14197–14201; published ahead of print November 2, 2015, http://dx.doi.org/10.1073/pnas.1517574112.

[Lan61] R. Landauer, "Irreversibility and heat generation in the computing process," *IBM Journal of Research and Development*, vol. 5, no. 3, July 1961, pp.183–191.

[Lan71] Bernard S. Landman and Roy L. Russo, "On a pin versus block relationship for partitions of logic graphs," *IEEE Transactions on Computers*, C-20(12), December 1971, pp. 1469–1479.

[Lee97] Thomas H. Lee, Mark G. Johnson, and Matthew P. Crowley, "Temperature sensor integral with microprocessor and methods of using same," U.S. Patent 5,961,215, October 5, 1999.

[Loh07] Gabriel H. Loh, Yuan Xie, and Bryan Black, "Processor design in 3D die-stacking technologies," *IEEE Micro, IEEE*, 27(3), May–June 2007, pp. 31–48.

[Lov16] Dominick Lovicott, *Thermal Design of the Dell PowerEdge T610, R610, and R710 Servers*, undated.

[May79] Timothy C. May and Murray H. Woods, "Alpha-particle-induced soft errors in dynamic memories," *IEEE Transactions on Electron Devices*, ED-26(1), January 1979, pp. 2–9.

[McW80] Thomas M. McWilliams, *Verification of Timing Constraints on Large Digital Systems*, Ph.D. Thesis, Stanford University, May 1980.

[Mea79] Carver Mead and Lynn Conway, *Introduction to VLSI Systems*, Addison-Wesley, 1979.

[Men97] Sunetra K. Mendis, Sabrina E. Kemeny, Russell C. Gee, Bedabrata Pain, Craig O. Staller, Quiesup Kim, and Eric R. Fossum, "CMOS active pixel image sensors for highly integrated imaging systems," *IEEE Journal of Solid-State Circuits*, 32(2), February 1997, pp. 187–197.

[Mil20] John M. Miller "Dependence of the input impedance of a three-electrode vacuum tube upon the load in the plate circuit," Scientific Papers of the Bureau of Standards, 15(351), 1920, pp. 367–385.

[Nie85] Niels J. Nielsen, "History of ThinkJet printhead development," *Hewlett-Packard Journal*, 36(5), May 1985, pp. 4–10.

[Nob56] Nobel Media AB 2014, "The Nobel Prize in Physics 1956," Nobelprize.org. http://www.nobelprize.org/nobel_prizes/physics/laureates/1956/.

[Nob73] Nobel Media AB 2014, "The Nobel Prize in Physics 1973," Nobelprize.org. http://www.nobelprize.org/nobel_prizes/physics/laureates/1973/josephson-facts.html.

[Nob00] Nobel Media AB 2014, "The Nobel Prize in Chemistry 2000," Nobelprize.org. http://www.nobelprize.org/nobel_prizes/chemistry/laureates/2000/.

[Nob00B] Nobel AB Media 2015, "The Nobel Prize in Physics 2000," Nobelprize.org. http://www.nobelprize.org/nobel_prizes/physics/laureates/2000/kilby-facts.html.

[Nob09] Nobel Media AB 2014, "The Nobel Prize in Physics 2009," Nobelprize.org. http://www.nobelprize.org/nobel_prizes/physics/laureates/2009/.

[Nob14] Nobel Media AB 2014, "The 2014 Nobel Prize in Physics — Press Release," Nobelprize.org. http://www.nobelprize.org/nobel_prizes/physics/laureates/2014/press.html.

[Noy61] Robert N. Noyce, "Semiconductor device-and-lead structure," U.S. Patent 2,981,877, April 25, 1961.

[Ono07] Shinya Ono, Koichi Miwa, Yuichi Maekawa, and Takatoshi Tsujimura, "V_T compensation circuit for AM OLED displays composed of two TFTs and one capacitor," *IEEE Transactions on Electron Devices*, 54(3), March 2007, pp. 462–467.

[Pan09] Panasonic, *Failure Mechanism of Semiconductor Devices*, T04007BE-3, April 2009.

[Par12] Hongsik Park, Ali Afzali, Shu-Jen Han, George S. Tulevsky, Aaron D. Franklin, Jerry Tersoff, James B. Hannon and WIlfried Haensch, "High-density integration of carbon nanotubes via chemical self-assembly," *Nature Nanotechnology*, 7, December 2012, pp. 787–791.

[Pav99] Paolo Pavan and Roberto Bez, "The Industry Standard Flash Memory Cell," Chapter 2 in Paulo Cappelletti, Carla Golla, Piero Olivo, and Enrico Zanoni, *Flash Memories*, Boston: Kluwer Academic Publishers, 1999.

[Pec97] Kim Peck, *PICmicro™ Microcontroller Oscillator Design Guide*, Microchip Technology, Inc., An588, DS00588B, 1997.

[Pow16] Powerstream.com, "Battery comparison chart—rechargeable," April 4, 2016, http://www.powerstream.com/Compare.html.

[Pre58] Ben Preece, "Flying high at zero altitude," *Modern Mechanix*, December, 1958, pp. 41–121, http://blog.modernmechanix.com/giant-analog-flight-simulator/.

[Pre13] John Preskill, "Quantum entanglement and quantum computing," *Caltech News*, http://www.caltech.edu/news/quantum-entanglement-and-quantum-computing-39090.

[PTM15] Arizona State University, Predictive Technology Model, http://ptm.asu.edu.

[Ram65] Simon Ramo, John R. Whinnery, and Theodore van Duzer, *Fields and Waves in Communication Electronics*, New York: John Wiley and Sons, 1965.

[Roy00] The Royal Swedish Academy of Sciences, "The Nobel Prize in Chemistry, 2000: Conductive polymers," 2000. http://www.nobelprize.org/nobel_prizes/chemistry/laureates/2000/advanced-chemistryprize2000.pdf.

[Rub83] Jorge Rubinstein, Paul Penfield, Jr., and Mark A. Horowitz, "Signal delay in RC tree networks," *IEEE Transactions on CAD*, CAD-2(3), July 1983, pp. 202–211.

[Sak91] Takayasu Sakurai and A. Richard Newton, "Delay analysis of series connected MOSFET circuits," *IEEE Journal of Solid-State Circuits*, 26(2), February 1991, pp. 122–131.

[Sak91B] Takayasu Sakurai and A. Richard Newton, "A simple MOSFET model for circuit analysis," *IEEE Transaction on Electron Devices*, 38(4), April 1991, pp.887–894.

[Sal12] Emore Salman and Eby G. Friedman, *High Performance Integrated Circuit Design*, New York: McGraw-Hill, 2012.

[Sch13] Derek K. Schaeffer, "MEMS inertial sensors: a tutorial overview," *IEEE Communications Magazine*, April 2013, pp. 100–109.

[Shu13] M. Shulaker, G. Hills, N. Patil, H. Wei, H. Chen, G. Gielen, G., and S. Mitra, "Carbon nanotube computer," *Nature*, 501(7468), 2013, pp. 526–530.

[Sea16] Sears, *Craftsman Arc Welder*, http://www.sears.com/craftsman-arc-welder/p-00920566000P?prdNo=1&blockNo=1&blockType=G1.

[Seq75] Carlo H. Séquin and Michael F. Tompsett, *Charge Transfer Devices*, New York: Academic Press, 1975.

[Ser07] Dimitrios N. Serpanos and Wayne Wolf, "VLSI models of network-on-chip interconnect," In: *IFIP International Conference on Very Large Scale Integration, 2007. VLSI – SoC 2007*, October 15–17, 2007, pp. 72–77.

[Ses64] Gerhard M. Sessler and James E. West, *Electroacoustic transducer,* U.S. Patent 3,118,022, January 14, 1964.

[She98] Kenneth L. Shepard and Vinod Narayanan, "Conquering noise in deep-submicron digital ICs," *Design & Test of Computers, IEEE*, 15(1), January–March 1998, pp. 51–62.

[Sho88] Masakazu Shoji, *CMOS Digital Circuit Technology*, Englewood Cliffs NJ: Prentice Hall, 1988.

[Sie82] Daniel P. Siewiorek and Robert S. Swarz, *The Theory and Practice of Reliable System Design,* Digital Press, 1982.

[Ska04] Kevin Skadron, Mircea R. Stan, Karthik Sankaranarayanan, Wei Huang, Sivakumar Velusamy, and David Tarjan, "Temperature-aware microarchitecture: modeling and implementation," *ACM Trans. Archit. Code Optim.* 1(1), March 2004, pp. 94–125.

[Ske98] Kenneth D. Skeldon, Lindsay M. Reid, Vivienne McInally, Brendan Dougan, and Craig Fulton, "Physics of the Theremin," *American Journal of Physics*, 66(11), November 1998, pp. 945–955.

[Smi85] F. M. Smits, ed. *A History of Engineering and Science in the Bell System: Electronics Technology (1925–1975)*, AT&T Bell Laboratories, 1985.

[Smo04] J. A. Smolin, "The early days of experimental quantum cryptography," In: *IBM Journal of Research and Development*, 48, no.1, January 2004, pp .47–52.

[Sze81] S. M. Sze, *Physics of Semiconductor Devices*, second edition, New York: John Wiley and Sons, 1981.

[Swa60] J. A. Swanson, "Physical versus logical coupling in memory systems," In: *IBM Journal of Research and Development*, 4(3), July 1960, pp.305–310.

[Tau97] Yuan Taur and Tak h.Ning, Fundamentals of Modern VLSI Devices, Cambridge: Cambridge University Press.

[Tof81] Tomasso Toffoli, "Bicontinuous extensions of invertible combinatorial functions," Mathematical Systems Theory, 14, 1981, pp. 13–23.

[Tra15] Wikipedia, "Transistor count," https://en.wikipedia.org/wiki/Transistor_count.

[Wal77] Jearl Walker, "The Amateur Scientist: Wonders of physics that can be found in a cup of coffee or tea," *Scientific American*, November 1977.

[Wik16] Wikipedia, "Energy density," http://www.wikipedia.org/wiki/Energy_density.

Index

'*Note*: Page numbers followed by "f" indicate figures and "t" indicate tables.'

A

Accelerometers, 218—220
Acceptor, 26
Activation energy, 199
Active matrix, 207—208
Active pixel sensor (APS), 214
Acyclic network, 101
Aggressor wire, 125
Aging effects, 152
Alternative structures, 47—48
Amdahl's law, 161, 162f
Analog computers, 7—9
Analytical engine, 4
Anode, 186
Arrhenius' equation, 199
Arrhenius prefactor, 199
Atanasoff-Berry computer, 6f

B

Babbage machines, 3—4
Ballistic, 222
Bandgap, 24
    reference, 200
Bands
    conduction band, 19
    defined, 19
    valence band, 19
Barriers, 94, 94f
Base states, 225
Bathtub curve, 152, 153f
Batteries, 185—188
Binary coding, 6—7
Binary logic, 6—7, 6f
Bit line, 169—170
Boltzmann's constant, 22—23
Boolean algebra, 4
Branching factor, 164—165
Bucket brigade, 213
Built-in potential, 30—31
Bus, 155—159

C

Cache/clock ratio, 174
Cache hit, 172—173
Cache miss, 172—173

Capacitive charge divider, 170
Capacitive coupling, 124f—125f
Capacitive load, 77, 77f
Carbon nanotubes, 221—223
    nanotube transistors, 222—223, 222f
Carrier concentrations, 26—27
Cathode, 186
Central processing unit (CPU), 7
    busses, 155—159
    clocking, 161—167
    electronic computers, 7
    global communication, 159—161
    interconnect, 155—159
    microprocessor characteristics, 153—155
Channel, 37—38
Charge-coupled device (CCD), 212
Circuits
    basic analysis, 235—237
    differential equations, 238—239
    effects, 111, 112f
    Kirchhoff's laws, 234—235
    ladder networks, 237
    linear time-invariant systems, 239
    models, 234
    RLC device laws, 233—234
    series and parallel networks, 236
    voltage dividers, 236
Clean rooms, 52
Clocking, 134—140, 161—167
    domains, 140—141, 166, 166f
    period, 135
    skew, 138
Combinational logic networks, 100
    delay and power, 108—109
    discrete values, 101f
    event model, 99—100
    gain
        delay, 105—108
        reliability, 103—104
    network model, 100—103
    noise and input/output coupling, 115—116, 115f
    noise and reliability, 109
    power supply and reliability, 109—115
Complementary MOS (CMOS), 64—67, 64f
    carbon nanotubes, 221—223
    imager, 214
    quantum computers, 223—228

Computational lithography, 56—57
Conditional probability, 241
Conduction band, 19
Conductivity, 22
Control
    mechanical computing devices, 2—4
Critical paths, 102, 102f
Crosstalk, 124—126
Cumulative distribution function (CDF), 241
Cumulative failure rate, 150
Current, 27
    leakage, 45
    subthreshold, 45

D
Decoupling capacitance, 113, 114b
Delay, 101—102, 105—108
    drive, 77, 83—84, 83f
    formulas, 78
    loads, 77, 83—84, 83f
    power, 108—109
    RC models, 76—83, 76f, 78f
    transistor models, 73—76, 73f
Demand paging, 181
Depletion region, 30—31
Device model, 38
    derivation, 40
    summary, 43
Difference machine, 3—4
Diffusion, 27
    current, 28, 28f
Digital light processor (DLP), 209
Discrete memory, 2—3
Displays, 205—211
Distributed model, 118
DNA reversible computers, 225
Donor materials, 26
Doping, 26
Double-patterning, 56—57
Drain, 37—38
Drain-induced barrier lowering (DIBL), 46—47
DRAM/cache ratio, 174
Drift, 20, 27
Drift velocity, 20
Drive, 77, 83—84
Driver chains, 84
DVFS. *See* Dynamic voltage and frequency scaling (DVFS)
Dynamic energy, 88
Dynamic RAM (DRAM), 169, 170f
Dynamic voltage and frequency scaling (DVFS), 89, 188—189

E
Earthed grounds, 65
Edison Effect, 13—15, 14f
Effective resistance, 73
Electret microphone, 218
Electromigration, 199
Electronic, binary, turing machines (EBT), 1
Electronic computers
    computer system metrics, 9—10
    defined, 1, 6—9
    mechanical computing devices, 2—4
    theories of computing, 4—6
Electronic link, 210
Electrostatics, 205
Elmore delay, 122, 123b
Elmore model, 122, 122f
Energy, 63
    barriers, 94
    dynamic energy, 88
    short circuit current, 86, 87f
    specific, 186
    switching energy, 85
Entanglement, 227
Error correction codes (ECCs), 176—177, 176f
Error rates, 95
    device and system, 152
Errors, 95
Event model, 99—100, 100f
Exponential distribution, 242
Exponentially tapered, 84
Extrinsic silicon, 26

F
Failure rate, 150
Fall time, 79—80
Fanin(n) function, 102
Fermi level, 25
Fill factor, 212
Flash memory, 178—181, 179f
Flatband, 35—36
Fleming valve, 15, 16f
Flip-flops, 132
Forward bias, 31
Fourier's Law of Heat Conduction, 192—193
Fringing capacitance, 118
Fullerene, 221—222
Functional testing, 53
Fundamental limits, 95

G

Gain, 103–108, 104f
Gap energy, 24–25
Gate delay models, 100
Gate delay scaling, 91, 91f
Gate topologies, 244–246
Gaussian distribution, 242
Glitches, 100
Global communication, 159–161
Globally asynchronous locally synchronous (GALS), 166
Ground, 65
  bounce, 111

H

Hamiltonian, 225
Hazard function, 151
Heat reliability, 198–200
Heat sink, 193–194, 193f
Heat transfer
  characteristics, 190–192
  defined, 189–200
  heat and reliability, 198–200
  modeling, 192–198
  thermal management, 200
Heterogeneous architectures, 182–183
Holes, 25–26
Hollerith punch cards, 4
Horowitz slope-dependent delay model, 246

I

Image sensors, 211–216
Impedance, 234
  matching, 84
Imrefs, 27
Inertial delay, 100
Inertial sensors, 218–220, 219f
Infant mortality, 152
Inkjet printer, 210–211
Input don't-care, 66
Input/output devices (I/O devices)
  accelerometers, 218–220
  displays, 205–211
  image sensors, 211–216
  inertial sensors, 218–220, 219f
  microphones, 217–218
  overview, 205
  touch sensors, 216–217
Integrated circuit (IC)
  defined, 48–59
  electronic circuits, 13–18

lithography, 55–57
manufacturing processes, 51–54, 51f–52f
Moore's Law, 49–51, 50f
separation of concerns, 58–59
vacuum tube devices, 13–15
vacuum tube triode, 15–18
yield, 57–58
Interconnect (IC), 155–159
  crosstalk, 124–126
  parasitic impedance, 116–118
  Rent's rule, 126–128
  transmission lines, 118–123, 119f–120f
  wiring complexity, 126–128, 126f
International Technology Roadmap for Semiconductors (ITRS), 50–51
Intrinsic material, 26
I/O devices. *See* Input/output devices (I/O devices)

J

Jacquard loom, 2–4, 3f
Josephson effect, 226–227
Junction, 30
Junction capacitance, 118

K

Kirchhoff's Current Law (KCL), 234, 235f
Kirchhoff's laws, 234–235
Kirchhoff's Voltage Law (KVL), 234, 235f

L

Ladder networks, 237
Lambda calculus, 4
Large-scale transmission lines, 119
Latches, 132
Latency, 137
Leakage control, 89
Leakage current, 45
Linear, 40
Linear amplifiers, 15
Linear region, 38–39
Linear time-invariant (LTI) system, 239
Liquid crystal display, 206, 207f
Lithography, 55–57
Loads, 77, 83–84
Logic gates
  complementary MOS, 64–67, 64f–65f
  delay, 72–84
  energy, 63, 85–90
  gate topologies, 244–246

Logic gates (*Continued*)
  Horowitz slope-dependent delay model, 246
  performance, 63
  power, 85—90
  reliability, 63—64, 94—96
  Sakurai—Newton model, 246—248
  scaling theory, 90—94
  static gate characteristics, 67—72
Logic levels, finding, 69
Long-channel devices, 46—47
Long-channel model, 38

M
Magnetic core memory, 7
Magnetic disk drives, 177—178, 177f
Mask, 52
Mass storage, 177—182
  flash memory, 178—181, 179f
  magnetic disk drives, 177—178, 177f
  storage and performance, 181—182
Materials
  Boltzmann's constant, 22—24
  defined, 18—29
  donor, 26
  intrinsic material, 26
  metals, 19—22
  semiconductors, 24—29
  temperature, 22—24
Mathematical logic, 4
Mechanical governor, 2, 2f
Memory
  defined, 168—177
  discrete, 2—3
  discrete memory, 2—3
  DRAM reliability, 176—177
  DRAM systems, 174—175
  magnetic core, 7, 8f
  structures, 168—172
  system performance, 172—174, 173f
MEMS. *See* Microelectricalmechanical system (MEMS)
Metal-oxide semiconductor (MOS), 33—37
Metal wire resistance, 110b
Metastability, 140—144
  failures, 144b
  modeling, 142
Microelectricalmechanical system (MEMS), 205, 218
Microphones, 217—218
Microprocessor
  characteristics, 153—155
Middle voltage, 71, 72f
Miller effect, 115

Mobile systems, 185—188
Mobility, 19
Moog synthesizer, 7—9, 9f
Moore's Law, 46—47, 49—51, 50f, 83—84
MOS capacitor, 33—37
  operation, 35—36, 35f
  structure, 34f
  technology trends, 35—36
  values, 34—35
MOSFET, 243
MOS field-effect transistor (MOSFET)
  advanced characteristics, 45—48
  basic operation, 37—45
  structure, 37
Multilevel cell (MLC), 180, 180f
Multiprocessor systems-on-chips, 182—183

N
Nanotube computer, 223
Nanotube logic, structures, 222
Nanotube transistors, 222—223, 222f
Negative feedback, 2
Netlist, 101
Noise and input/output coupling, 115—116
Noise margin, 69
Nominal parameter, 57
Nonlinear amplifiers, 15
n-type transistor, 37—38

O
Ohm's Law, 233
One-sided timing constraint, 135—136
Organic LED, 208
Output don't-care, 66
Output function, 130

P
Pads, 53—54
Page fault, 181
Parallelism, 102
Parametric failures, 57
Parametric testing, 53
Parasitic impedance, 116—118
Passive matrix, 207—208
Patent, 13—15
Permanent fault, 130—131
Permanent faults, 139—140
Peukert Effect, 188
Phase-locked loop (PLL), 167
Photodetection, 211

Photodiode, 211–212
Photogate, 214
Phototransistor, 211–212
Pierce oscillator, 167, 167f
Piezoelectric effect, 166–167, 167f
Pinchoff, 42–43
Pin inductance, 110–111
Pins, 53–54
Pipelining, 136, 137f
Planar processing, 51–52
PLAs. *See* Programmable logic arrays (PLAs)
p-n junctions, 30, 30f, 243
Poisson distribution, 242
Pool model, 78, 79f
Power, 88
    consumption, 95, 182–189
        mobile systems and batteries, 185–188
        power management, 188–189
        server systems, 182–185
    delay, 88, 108–109
    density, 92
    leakage control, 89
    management, 188–189
    race-to-dark, 89–90
    reliability, 109–115
    scaling, 92
    static power, 88
    total power, 88
Prefetching, 173
Primary inputs, 100
Primary outputs, 100
Probability, 241–242
Probability density function (PDF), 241
Probability mass function (PMF), 241
Process corners, 57
Processor pinout, 111b
Processors, 153–167
Process variability, 57
Programmable logic arrays (PLAs), 222–223
Proof mass, 218
p-type transistor, 37–38
Pulldown transistor, 64–65
Pullup transistor, 64–65

Q
Quantum computers, 223–228
Quantum cryptography, 227
Quantum efficiency, 206
Quantum reversible computers, 225
Quasi-Fermi levels, 27
Qubit, 225

R
Race-to-dark (RTD), 89–90, 189
Random-access memory (RAM), 168–169
Random walk, 19
Rayleigh's criterion, 55–56, 55f
RC models, 76–83
Registers
    dynamic register, 131
    static register, 131
Register timing, 133
Register type, 137
Reliability, 10, 63–64
    barriers, 94, 94f
    errors, 95
    fundamental limits, 95
    heat, 198–200
    noise and, 109
    power supply, 109–115
Rent's constant, 128
Rent's rule, 126–128
Resistivity, 22
Restoring logic values, 69–70
Reverse bias, 31
Reversible computation, 224
Rise time, 79–80
RLC device laws, 233–234

S
Sakurai–Newton model, 246–248
Saturating inverting amplifier, 103–104
Saturating logic, 6–7, 69–70
Saturation, 40
    region, 38–39
Scaling, 90–94
    gate delay scaling, 91, 91f
    model, 90–91, 90f
    power scaling, 92
    scaling model, 90–91, 90f
    wire scaling, 92, 92f
Semiconductor diode, 30–33
Sequential machines, 129, 129f
    clocking, 134–140, 135f–136f
    combinational logic, 99–116
    interconnect, 116–128
    metastability, 140–144, 141f–142f
    registers, 131–134
    sequential models, 128–131
    timing, 134–135
Server systems, 182–185
Setup/hold times, 136
Sheet resistance, 117

Shielded wires, 125f
Short-channel effects, 243
Short circuit current, 86, 87f
Simulation, 103
Solid-state devices, 29—48
Solid-state disk (SSD), 181
Solid-state physics, 24
Source, 37—38, 101
Species, 188
Specific energy, 186
Speed—power product, 88
Spring constant, 218
State transition, 130
    table, 130—131, 130f
Static curves, 67
Static gates, 67—72
Static power, 88
Static property, 67
Steady state, 194
Storage
    performance, 181—182
Substrate, 33
Subthreshold current, 45, 46f
Subthreshold slope, 45
Subthreshold swing, 45
Superposition, 239
Surface potential, 36
Switching energy, 85
Synchronizers, 166
System power consumption, 182—189
System reliability, 150—152

T

Technology node, 50
Telegrapher's equations, 119
Temperature, 23, 24f, 47
Thermal capacitance, 191
Thermal management, 200
Thermal resistance, 191
Thermionics, 13—15
Thin film transistor, 207—208
Threshold potential, 36
Throughput, 137
Through-silicon vias (TSVs), 175
Time constant, 79
Total power, 88

Touch sensors, 216—217
Transconductance, 15, 43
Transfer curves, 67
Transient analysis, 194
1-transistor (1T DRAM), 169—170
Transistor behavior, 106—107
Transistor models
    defined, 73—76, 73f
    resistive approximation, 74f
Transistor sizing, 82
Transition, 99
    time, 79—80
Transmission lines, 118—123, 119f
    propagation, 120
Transport delay, 100
Triple-level cell (TLC), 180
6T SRAM, 171—172, 171f
Tube amplifiers, 15
Turing machines, 1, 5, 5f
Twin-tub processing, 64
Twizzled wires, 126f
Two-phase machine, 138
Two-sided timing constraint, 138

U

Unit capacitance, 117—118

V

Vacuum tube diode, 15
Valence band, 19
Victim wire, 125
Virtual memory system, 181
III-V material, 206
Voltage dividers, 236
von Neumann machine, 7, 8f

W

Waveforms, 76
Wire length estimation, 127—128
Wire scaling, 92, 92f
Wiring complexity, 126—128
Word line, 169—170
Worst-case delay, 102

Printed in the United States
By Bookmasters